Memory and Dreams

MEMORY AND DREAMS
The Creative Human Mind

George Christos

Rutgers University Press
New Brunswick, New Jersey, and London

British Cataloging-in-Publication information is available from the British Library.

Copyright © 2003 by George Christos
All rights reserved
No part of this book may be reproduced or utilized in any form or by any means, electronic or mechanical, or by any information storage and retrieval system, without written permission from the publisher. Please contact Rutgers University Press, 100 Joyce Kilmer Avenue, Piscataway, NJ 08854–8099. The only exception to this prohibition is "fair use" as defined by U.S. copyright law.

Design by John Romer

Manufactured in the United States of America

Library of Congress Cataloging-in-Publication Data

Christos, George.
    Memory and dreams : the creative human mind/George Christos.
       p. cm.
    Includes bibliographical references and index.
    ISBN 0-8135-3130-6 (alk. paper)
    1. Memory. 2. Dreams. 3. Sudden infant death syndrome. I. Title.

QP406.C477   2003
153.1′2—dc21   2002070504

This book is dedicated to Daniel, Kelly, Johnny, and Jenny.

IN MEMORY OF DR. EVAN OWEN (1933–2000).

Contents

PREFACE ix
ACKNOWLEDGMENTS xiii
1. Introduction and Overview 1
2. The Electrochemical Brain 11
3. The Remembering Brain 37
4. The Creative Brain 72
5. The Dreaming Brain 105
6. Unraveling the Mystery of Sudden Infant Death Syndrome (SIDS) 156
    REFERENCES 199
    INDEX 219

Preface

The main purposes of this book are to present a tenable theory of how memory is stored, processed, retrieved, and manipulated in the brain; to put forward ideas of how the brain can generate novel information and creative ideas; to contemplate what the brain may be doing during the particularly active phase of sleep associated with dreaming; and to deliver my theory about the cause of sudden infant death syndrome (SIDS) to the wider public and scientific communities. I have tried to keep the discussion at a level that is accessible to the general reader, but I hope it is also valuable to other scientists.

My interest in brain function started in 1991, when I first saw the amazing book *Modeling Brain Function,* by Daniel Amit (1989). I was fascinated to learn how researchers like John Hopfield and Amit set about quantifying the main attributes of memory in terms of simple concepts and mathematical models. In 1991 I attended the Second Australian Conference on Neural Networks, where I heard Bill Gibson of Sydney University talk about the Crick-Mitchison hypothesis that dreaming may involve a reverse-learning (or unlearning) mechanism that helps the brain to eliminate unwanted spurious memory. This started my interest in understanding the function of rapid-eye-movement (REM) sleep, the phase of sleep during which dreaming occurs. In 1992 I started conducting computer simulation experiments on neural network models to test the Crick-Mitchison hypothesis. I decided around this time to broaden my knowledge and find out what other people, particularly psychologists (like Sigmund Freud), neuroscientists (like Allan Hobson), and philosophers, thought about dreaming. This started my journey away from the purer mathematical sciences.

Also in 1992, I met a commuter on the morning train, Mark Peaty, who worked an office job in Perth but had a keen layman's interest in brain function, particularly with consciousness. Peaty first brought to my attention the subject of "lucid dreaming," a state in which a sleeper actually becomes conscious while dreaming. From that point on, the sleeper is generally able to exert some control over the dream content. Researchers at the Stanford University Sleep Research Center, headed by Steven LaBerge, have used lucid dreaming to show that when we dream, our bodies try to act out the content of our dreams as if they were real.

In August 1992, at a dinner party at my brother's house, I was telling everyone about the fascinating subject of lucid dreaming and how LaBerge wanted to find out whether he actually held his breath when he dreamed he was swimming underwater. The key to his investigation was that although the body is practically disconnected from the brain during dream sleep, certain nerve fibers and attached muscles remain active, such as those that control the characteristic eye movements, the lungs, and the heart. The Stanford researchers used these partially active channels to communicate during a lucid dream to their colleagues in the laboratory. Typically a sleeper used unconventional up-and-down eye movements, as opposed to the sideways eye movements usually associated with REM sleep, to signal the beginning of a lucid dream. The sleeper's colleagues would then start to monitor the body of the sleeper and record physiological activity. On awakening, the dreamer would write a report, and this would be compared to the laboratory recording, which revealed that the sleeper did indeed hold his breath while he dreamed that he was holding his breath. This also appeared to last as long as it was perceived to last in the dream.

At the dinner party, the subject of lucid dreaming and holding one's breath underwater naturally led to the subject of water babies and how they can hold their breath underwater for very long periods of time. This was the moment that the idea behind my SIDS theory was germinated. I knew that we dream about our own memory, about our past, and that a baby would necessarily dream about its own memory, its life in the womb. In the course of that dream, since a fetus does not need to breathe (because the mother supplies it with oxygen through the blood), the baby might act out the dream, stop breathing, and consequently die. Just like the researcher who held his breath while he was dreaming that he was underwater, the baby would stop breathing while dreaming that it was back in the womb. This is the essence of my SIDS theory, which to a large extent was initiated by comments made by my wife (Christos 1992; Christos and Christos 1993; Christos 1995a). The theory is explained in detail in chapter 6, and lucid dreaming is discussed further at the end of chapter 5.

My simple hypothesis offers a possible explanation for the cause and the trigger of SIDS, and is consistent with almost all of the known facts about it. Most other SIDS theories are consistent with only one or two of the known facts and cannot explain what causes this tragic event. Some theories are based on only a single lifestyle issue (like smoking), which, although it may be correlated with SIDS, is not present in all cases of SIDS. I am quite encouraged that nine years after I first proposed my theory it is still viable. Some would suggest that this is because my theory cannot be tested, but this is not entirely true. It can be tested (and has been) against the known facts about SIDS, and I have proposed new ways in which it can be tested further. Some recent developments, such as the reduced risk associated with the use of baby pacifiers (called "dummies" in England and Australia), support my theory.

This book, however, is not just about SIDS, but is mainly about how memory works in the brain, how it is stored, retrieved, combined, manipulated, and maintained; about creativity; and about dream sleep. Chapters 2, 3, and 4 are about memory and how the brain works, and chapter 5 is about how dreaming aids in the reprocessing of memory. The nuts and bolts of how the brain works at the microscopic level are described in chapter 2. Chapter 3 deals with memory, or the recollection of past experiences, and how it works in simple neural network models and the brain. Where is memory stored in the brain, and how is it retrieved?

One of the main conclusions about the way memory is stored in the brain is that it is distributed and individual memories overlap each other. A natural consequence of this type of storage of memory is that the brain generates its own sets of memories, so-called spurious memories. These spurious memories are generally considered to be a nuisance, and much work in the past has been directed toward finding ways to eliminate or to control them. In the original version of this book, written two years earlier, the Crick-Mitchison unlearning hypothesis was one of the main topics. The basic idea behind this hypothesis is that the brain is throwing away spurious memories during dream sleep, or unlearning them. My viewpoint toward spurious memories has changed considerably while I have been rewriting this book, although I have always appreciated that spurious memories may be of some importance. I now firmly believe not only that spurious memories are useful, but that they may be essential. Spurious memories may be the basis of creativity (or new ideas). They may be required to learn something new. Without them we would just recall and relearn what we already know. Spurious memories may be needed to allow us to make (small) mistakes and to adapt (or evolve our knowledge). Without spurious memories it is difficult to make mistakes in neural networks, as neural networks are naturally noise and error tolerant. Spurious memories may also allow us to generalize, and to make associations between memories, by combining similar memories in novel ways. This is a profound shift in the general philosophy of such phenomena and forms one of the main theses of this book. These matters are considered in chapter 4.

The other main subject of this book is the function of dream sleep. It is quite remarkable that we know so much about so many things but we still do not know why we dream, or even why we sleep for that matter. What makes dream sleep particularly interesting is that the brain is extremely active during this phase of sleep, and almost all other mammals dream as well. Therefore, it would seem to serve some important biological function. Most theorists believe that the brain is engaged in some sort of internal processing of memory during dream sleep. This makes sense because to a large extent the brain is disconnected from the outside world during dreaming. In writing this book I have drawn together various theories about the function of dream sleep and devised a few new theories of my own along the way. Could it be associated with throwing away spurious memory, as suggested by Crick and Mitchison, or could it be associated with generating more spurious states to prepare us for a new day of learning? Ultimately the only way to decide among these theories is by experimentation and computer simulation.

Each of the chapters in this book is relatively self-contained and does not rely too heavily on material from the previous chapters. By the same token, there is some repetition in the various chapters. Readers who are primarily interested in SIDS can go directly to chapter 6, but I hope they will read what else I have to say about memory and dream sleep in general. Chapter 1 acts as an introduction and summary to what this book is about.

Finally, I would like to apologize to anyone whose work I have not properly referenced, or whom I may have misrepresented by my misinterpretation of their ideas.

Acknowledgments

I would like to thank my wife, Jenny, for her encouragement and for reading the entire manuscript, my colleague and friend Jamie Simpson for his encouragement, Richard Horsley for his support, Francis Crick for discussions about dream sleep, and Vicki Schechtman and Warren Guntheroth for discussions about sudden infant death syndrome. I would like to express my deepest gratitude to Evan Owen (my specialist doctor) and Nigel Bill (my family doctor) for helping me through a rather difficult period of ill health. Thanks to Carey Ryken-Rapp, and Judy Wheeler for reading parts of the manuscript, Anita Littlewood for doing some of the sketches for me, and Arnold Scheibel and George Paxinos for providing photographic material. A special thank you to all the staff at Rutgers University Press, and in particular Helen Hsu, Audra Wolfe, Marlie Wasserman, and Stephen Mayer (copyeditor), for helping me prepare the manuscript. I should also thank a number of other people, unmentioned for obvious reasons, whose skepticism and negativity actually spurred me on.

MEMORY AND DREAMS

CHAPTER 1

Introduction and Overview

The human brain and the human mind, which are probably one and the same, have fascinated scientists, philosophers, and most of the rest of humankind for centuries, perhaps much longer. The human brain is a truly remarkable organ that is responsible for our abilities to move and control our bodies and organs; to process visual, auditory, and other sensory information; to perceive; to react; to experience emotions; to remember; to think; to make decisions; to plan; to communicate with each other (by various means such as speech or the written word); to act in a social and responsible manner; and to understand complex phenomena, like scientific theories and the brain itself. The human brain also gives us the sensation that there is a little person inside us, or a "self," who is in control of what we are doing. What is even more remarkable is that the brain is only the size of a small coconut (or a decent sized mango fruit), which you can hold in the palm of your hand, and weighs about 1.5 kilograms (or 3 pounds). In appearance the brain looks like a gray, white, and mushroom-colored overgrown walnut, with large wrinkles and numerous networks of blood vessels on its surface. It has the consistency of a jelly or porridge, and falls apart if taken out of the human skull (see figure 1.1).

To really appreciate the splendor of the human brain, the next time you go out into the street or to a shopping center, stop for a moment and observe the people walking around you, talking to each other, avoiding collisions with each other, running businesses, buying goods, driving cars, and flying airplanes. The human brain controls all of these functions and activities, and what is more, the human-created objects around you, like buildings, cars, and roads, the rich and sophisticated human culture and society, and the vast technology have all resulted from the interaction among human minds. On top of all of these things, every person is a conscious, thinking, and planning human being, living in a complex world.

Figure 1.1. Dr. George Paxinos of the University of New South Wales, Sydney, holds a human brain in the palm of his hand and marvels at how this simple-looking organ is capable of producing so many amazing functions. *(Photograph courtesy of George Paxinos.)*

Although the human brain is exceptional in its capabilities, even the brains of animals and insects are quite amazing. The brain of an ant is the size of a grain of sand, yet this enables the ant to move around (quickly) over complicated landscapes, pick up chemical ant-made scents, detect changes in temperature, air movement, and vibrations, search for food to take back to the nest, and interact socially with its own colony. The ability of insects like the bee to fly is also controlled by a tiny nervous system.

Considering the vast and astonishing capabilities of the human brain, it is not surprising that many scientists would assert that it is the most complicated system in the universe. Physicists and mathematicians cannot solve exactly the so-called three-body problem, where there are just three interacting particles. "Chaos" can evolve from such a simple system. In the human brain, there are on the order of a hundred billion to a thousand billion neurons (the basic building blocks of the brain), which are intricately connected and interact strongly with each other. These neurons communicate through small pulses of electricity and chemical currents. Each neuron simultaneously receives and interprets input from thousands to tens of thousands of other neurons, and based on this information decides whether it should itself "fire"—that is, emit an electrical pulse that travels to other neurons—or remain quiescent. Basically if a neuron receives sufficient excitation from other neurons, counterbalanced by neurons that act on it in an inhibitory role, it will fire.

Each individual neuron performs a very simple and almost trivial function, yet, as a whole, a collection of neurons is able to perform very complex tasks and functions, as outlined above. Brain function is an emergent collective property of a large number of neurons, which seems to encompass more than the sum of its parts. This sort of emergent behavior is a general feature of other large physical, biological, and social systems. Nature abounds with many beautiful patterns, structures, and systems that emerge from rather trivial local interactions. Just look at the way that cells in your body organize themselves to form a complex living creature, how water molecules are organized into clouds, and how society is organized, with its complex legal and financial systems.

In the brain, memory is a representation of the past. Clearly memory serves an important biological function as it gives animals a survival edge in that they can use previous knowledge to better acquire food, or to prevent a certain dangerous situation from reoccurring. In some respects, the way that memory is stored in the brain is similar (but not precisely equivalent) to how flowing water cuts out a network of trenches and rivers in a landscape. When it rains, the water will preferably flow along the same paths that were previously scoured out of the land. In the brain, when a particular channel between two neurons is used often, this channel normally becomes "enlarged" so that it is even more accessible in the future, and conversely if a particular channel is not used, it diminishes in its capacity. This is how learning takes place in a nervous system. By directing the flow of electricity through the same channels as before, the brain is able to reinitiate (or recall) the same electrical patterns, or memory states, that caused this change in the first place. In the human brain there are on the order of one million billion connections between neurons that can be varied in this way. The flow of information (or electrical current) across these junctions, where two neurons (almost) touch each other (called synaptic clefts), is controlled by the transfer of chemicals (called neurotransmitters). The amount of chemical current that is transferred across a synaptic gap is variable, and this is how memory is actually stored in the brain.

(The brain also has a type of memory referred to as "habituation," a process by which the nervous system becomes familiar with a stimulus and responds less often and less vigorously when it is repeatedly and persistently stimulated. In this case, stimulus-receiving sensory neurons actually release less neurotransmitters upon repeated firing, and have a diminished effect on response-output motor neurons. Repeated use of the same channel causes it to "shrink" instead of enlarge.)

A memory is represented by a certain stabilized pattern of firing neurons. When the brain receives a meaningful input, it processes the input into a stationary state (also called an attractor), where the activation states of the neurons collectively stabilize, or persist in the same state of excitation (or quiescence). This is how memory is recalled, or how we recognize that an input was familiar to us. Memories are not stored in any one particular neuron, but are distributed over a wide area of the brain involving many neurons, possibly many millions. Different memories may also share certain active neurons in their representations, so in this sense memories overlap each other. This is quite unlike the way memory is stored in a computer, where it has a unique address and a separate location in which it is stored. To retrieve a memory, a programmer just uses its address. In the brain, a memory is retrieved instead by the content of the input. If sufficient information or cues are provided, the brain will be able to retrieve that memory. This is a useful attribute of brain function because one can retrieve a memory with only part of its "address," whereas a computer would not respond if it was not given the entire address precisely. This explains why we are able to recognize people we "know," even though they may look quite different from when we last saw them. Because neurons work together, or collectively, the brain is quite impervious to errors in the input and to noise in general. If a small number of neurons are in the incorrect firing mode for a particular memory, the other neurons collectively correct those neurons. A computer, on the other hand, stops executing (what it was meant to do) if a single solitary instruction, or a single bit, is wrong.

It is fascinating to consider how a simple process of integrating and firing neurons can account for not only memory but a diverse range of brain functions such as language, conscious awareness, creativity, the ability to understand the world mathematically, and emotions like joy, sorrow, and anger, to mention just a few. Each neuron by itself is an unintelligent binary device ("on" or "off") in the overall machine, but somehow the neurons combine to generate these incredible properties of brain function. Over the last twenty or thirty years an incredible amount of progress has been made in understanding how the brain works, both experimentally and theoretically. Scientists now have a basic understanding of how memory is stored and retrieved in the brain, and how the brain is capable of doing some of the amazing things it can do. Much of what we know comes from studying simple mathematical models (called neural networks), examining brain-affected or brain-damaged patients, and conducting animal experiments.

In chapter 2, I give a brief description of the main components of the brain. Although the brain seems to be highly organized, there is insufficient information in our genes to tell it how to make all of its connections. The brain makes most of these connections through an evolutionary process that involves waves of rapid growth of random connections followed by periods of intense pruning. The pruning process keeps only the most important connections. How this actually happens is still a bit of a mystery. Most of this rapid development of the brain takes place during infancy and childhood, and coincides with the periods of most intense learning. Some aspects of learning, like the ability to see, must be developed during this critical period; otherwise one will never be able to see.

In chapter 3, we look at memory, the different types of memory, and the way that memory is stored in the brain. Memory is actually stored in the adjustable synaptic efficacies linking neurons, and is represented by patterns of active configuration in a number of different areas in the brain, involving modules or small networks of neurons. This storage arrangement facilitates the recall of memory from incomplete cues or partial information. One of the most important concepts explaining how memory and learning work in the brain (and possibly brain function in general) is that of an attractor, or a reverberation. A group of neurons collectively excite each other so as to temporarily remain in the same state of excitation. I will elucidate these concepts and demonstrate how "content-addressable memory" works in a simple neural network model, called the Hopfield model.

One of the problems with the way memories are stored in the brain, in a distributed and overlapping fashion, is that the brain also generates its own set of memory states, which are comprised of combinations of features of the stored memories. These so-called spurious memories were not intentionally stored in the brain. Simple mathematical neural network models suggest that without some sort of intervention, these spurious memories can reach catastrophic proportions, and the network will eventually be unable to retrieve any of the stored memories. For this reason, spurious memories, as their name implies, are generally considered to be a nuisance, and much theoretical work has been concerned with finding ways to limit their number and influence on intentionally stored memories.

With suitable control, spurious memories may, however, be a blessing in disguise. Humans are very creative, whereas computers are not, so where in the brain does creativity come from? Spurious memories are interesting candidates for creative states, as they are generally made up of combinations of features of stored memories and this is precisely what creative ideas are really like. Creative ideas generally combine different aspects of memories by putting bits and pieces of information together to arrive at a new way of doing something. They never come completely out of the blue. In chapter 4, we investigate the notion that these naturally occurring spurious states may correspond to creativity.

Spurious memories may also be useful for other purposes. If the brain simply processes input until it converges to attractors, or stored memory, how is it able to learn something new, and how does the brain make mistakes when it appears to be noise tolerant? I suggest that spurious memories may help resolve these problems by allowing the brain to learn new attractors and to adapt by making small mistakes. A network can make a mistake by selecting a spurious state that is similar to one of the stored memories. Because spurious states are made up of combinations of stored memories, they may also be important in making associations between memories. Other researchers have suggested that spurious memories may help the brain to generalize and categorize.

As neural networks generally converge to stored memories more rapidly than to spurious memories, one can distinguish among them. I use this fact in chapter 4 to propose an interesting theory for "déjà vu," the weird sensation that something is familiar to us when actually it is not. This may correspond to a situation where the brain converges quickly to a spurious memory and incorrectly identifies it as a stored memory, thinking it to be familiar. A similar argument, involving the slow convergence to a stored memory, could explain "jamais vu," the sensation that something is unfamiliar when it should be familiar.

Another remarkable thing about the brain is that it needs to sleep every day. Without sufficient sleep, we feel tired and are unable to concentrate, and the brain does not function properly. This in itself suggests that sleep is for the brain, and that it may involve some sort of internal processing. For some time it was thought that the main function of sleep was to help the body (and to a lesser extent the brain) to recover. Recuperation may well be one of the most important functions of sleep, but there are some periods during sleep when the brain is extremely active. During rapid-eye-movement (REM) sleep, the phase of sleep that is normally associated with dreaming, there is an increased amount of neural activity, and more blood flows into the brain during this phase of sleep than when we are awake. The electrical activity of the brain during REM sleep also looks remarkably similar to what it looks like when we are awake. What could the brain be doing, working so hard for hours every night while we sleep? As almost all mammals have evolved with REM sleep, it may serve some important biological function. The importance of REM sleep is also highlighted by the fact that if we are deprived of it on one night, we experience REM rebound, or catch-up REM, the next night.

During dream sleep the brain seems to be largely disconnected from the outside world and from its own body. The usual stimulation that it receives through its eyes, ears, and other sensory organs is replaced by what appears to be a random signal emanating from a small set of neurons (about ten thousand or so) located in the brain stem (the extension of the spinal cord into the brain). These signals find their way to the neocortex, which is the folded or wrinkled sheet of neurons that covers the top of the brain, where our memories and most intelligent processes reside, through the same channels that input from the outside world normally uses. The brain seems to process this noisy input from the brain stem as if it were real, and dreams are the result of the brain trying to make sense of this internal stimulation (Hobson and McCarley 1977). This explains why our dreams appear to be so real and yet so bizarre, and why they are mainly concerned with our own memories. The noisy nature of the stimulus suggests that dream content may be largely irrelevant, but on the other hand, dreams involve our memories and our emotions. The fact that the brain is largely disconnected from the outside world during REM sleep but is so active suggests that it may be engaged in some sort of internal processing, possibly of memory.

There is strong evidence to suggest that when we are dreaming of doing something (like singing) we are activating those neurons that we would use while actually doing this thing when we are awake. These neurons are not fully activated, however, if we are simply imagining to be performing this task. In this sense, dreaming is real. Fortunately there is another small group of neurons, also located in the brain stem, that prevent us from physically acting out our dreams. These neurons stop the brain signals from traveling down the spinal cord. This prevents us from hurting ourselves, or our partners, during sleep. When these deactivating neurons were removed from a cat's brain, the cat was observed to be acting out its dreams (stalking prey in the laboratory) while it was still in REM sleep.

In chapter 5, I will review what is known about REM sleep and describe some of the numerous theories about the possible functions of REM sleep and dreaming. Some theories suggest that it is concerned with the development of the brain, since infants and newborns have much more REM sleep than adults, but then why do adults continue to have REM sleep after the critical development period of the brain? Sigmund Freud, followed by other psychologists, thought that dreaming fulfils a psychological need, but this does not explain why infants and children have so much REM sleep. Freud was unaware of the fact that dreaming is associated with REM sleep and that infants have so much more REM sleep than adults. Why would an infant need so much more psychological relief? Other theories assert that dreaming is associated with the reprocessing, or consolidation, of memory. This can be achieved by unlearning the weak and unwanted spurious memories, or by rehearsing or relearning the strongest and more important memories. Another theory suggests that dreaming is a form of evening entertainment, like a television show, to stimulate the brain while the body rests.

Francis Crick (of DNA fame) and Graeme Mitchison suggest that during REM sleep the brain is engaged in a reverse-learning or unlearning process, predominantly affecting or eliminating spurious memories, which are more readily activated by the random noisy stimulation from the brain-stem neurons (Crick and Mitchison 1983). They have proposed that these unwanted states are eliminated or weakened by a process the opposite of learning. In terms of our waterways metaphor, the small creeks and channels are covered with a light sprinkling of sand. Computer simulations support the idea that this process helps to eliminate spurious memories and improve memory capacity. It also allows the brain to eliminate trivial memories acquired during the course of the day and to gradually forget other unwanted and irrelevant older memories. The reverse-learning idea explains why we are generally unable to recall our dreams, because we forget them as soon as we have had them. I will discuss this theory further, and many other theories on the function of REM sleep and dreaming, in chapter 5. Jonathan Winson, on the other hand, suggests that we are rehearsing or relearning important information during dream sleep (Winson 1985).

In more realistic neural network models, where spurious memories are not a problem and their number is limited and controlled, they may correspond to creativity and may be essential for the brain to learn new information, to think, and to adapt. With this in mind, I suggest a new theory for the function of dream sleep, which is concerned instead with the generation of more spurious memories. This added bumpiness on our memory store is what prepares us for a new day of learning. In models where the proportion of spurious memories is low, this process can take place via reverse learning.

The subject of "lucid dreaming," a form of dream sleep where sleepers actually become aware that they are dreaming, is briefly discussed in chapter 5. If they are able to remain asleep while they become lucid, they are generally also able to exert some control over the content of their dreams. At the same time, experienced lucid dreamers are able to communicate with their colleagues in sleep laboratory experiments. Although the brain is practically disconnected from the body during REM sleep, there are certain nerves and muscles that remain active, such as those that control the lungs, the heart, and associated eye movements. As mentioned in the preface, a group of researchers at Stanford University were able to use eye movements to tell colleagues in the laboratory that they were having a lucid dream (LaBerge 1986). The dreamers' report, on the subject of the dream, and the laboratory report, of recorded physiological activity, were later compared, and it was found that they indeed tried to act out their dreams as much as was possible. As an example, when a researcher dreamed that he was swimming underwater, he actually held his breath. Similarly, sexual dreams were associated with physiological signs of sexual activity, like vaginal contractions.

Infants have from five to eight hours of REM sleep each day during the first year of life, compared to only one or two hours each day for adults, so one may ask what are they dreaming about. They are dreaming about their own memories, which include memories of being back in the womb. There the fetus does not have to breathe, because the mother supplies it with oxygen through the blood. My hypothesis about sudden infant death syndrome (SIDS) is that, just as a sleeper dreaming of swimming underwater holds his breath, a baby dreaming of being back in the womb may also stop breathing, and die. Note that although the brain is practically disconnected from the body during REM sleep, important physiological functions remain active. Furthermore, the deactivation mechanism that normally stops us from acting out our dreams is not properly developed in a young infant, which makes it even more vulnerable.

SIDS is a mysterious phenomenon, with more than one in every thousand of all live births resulting in a death that seems to have no apparent cause. SIDS is diagnosed only after every other cause of death has been excluded, and after a rigorous investigation of the death scene and the family's medical history. In chapter 6, I explain in some detail how my simple theory fits in with all of the known facts about this mysterious and prevalent cause of infant death. One of the most remarkable findings about SIDS is that there are certain risk factors associated with it. For example, if a baby is put to sleep face-down on its stomach, its risk of SIDS is increased by a factor of three or more, compared to the face-up position. Knowing this has halved the number of SIDS cases, but there is no satisfactory explanation of why it is so, other than my theory. It has also been found that if an infant sleeps with a pacifier, its risk is decreased by a factor reportedly as much as twenty-fold. One cannot explain this by suggesting that pacifier sucking helps by keeping the airway passages open, because thumb sucking should then presumably have a similar effect, whereas it is actually associated with an increase in the risk of SIDS. My theory is consistent with the reduced risk associated with pacifier sucking and the increased risk associated with thumb sucking.

I suggest that, just as dreams can influence the body of the infant, the reverse is also true: the environment of the sleeping infant can influence its dreams. If the environment resembles the conditions in the womb, this would increase the risk of SIDS. When a baby is sleeping face-down, it generally assumes a semifetal position, tucking its arms and legs into its stomach. This may increase the possibility of a potentially fatal fetal dream. Thumb sucking is a risk factor because it is something that a fetus does in the womb, around seven months of gestation. Pacifier sucking, on the other hand, would have the opposite effect, reminding the infant that it has been born, that it is not in the womb, and that it needs to breathe. Another strange observation is that the risk of SIDS is reduced if an infant sleeps in the same room as adults but increases if the infant sleeps in the same bed. In other words, room-sharing reduces the risk of SIDS, whereas bed-sharing increases the risk of SIDS. I would suggest that if an infant sleeps too close to an adult, it may pick up the adult's heartbeat which may remind it of being back in the womb.

My simple theory is consistent with most, if not all, of the known facts about SIDS, whereas most other theories are consistent with only one or two of the known facts, and in some cases there is no other explanation.

CHAPTER 2

The Electrochemical Brain

The brain is an electrochemical machine that processes information from the environment, obtained through the senses. This information is processed according to the brain's previous experience (or memory) and generally results in some sort of appropriate action or response. The rapid nature of this response is what distinguishes animals, which have a brain and/or a nervous system, from plants. Plants generally respond slowly to changes in the environment and do not have true memory. For most animals this response is primarily concerned with survival, whereas in humans, the brain is a much more sophisticated device that has enabled them to be more reflective and to understand the world around them, including the brain itself. In the discussion that follows, I tend to include all sensory neurons, such as those located on the extremities of our bodies and in our eyes, as well as the neurons in the spinal cord as being part of the brain. It is more common, however, to call all of this the central nervous system (CNS), and the part of the CNS that resides in the head the brain. The distinction between the brain and the nervous system as a whole becomes quite muddied when we look at small animals.

Much of how the brain works is related to the collective behavior of billions of cells in it. In this sense the brain is a biological machine. An even better description would be to refer to it as a biological computer, although, as I have argued previously, it works quite differently from a conventional computer. (See also the discussion in subsequent chapters.) In this chapter I will give a broad description of how the various components of the brain work, how they are connected to each other, how they communicate with each other, and how they are organized. In the following chapters we will look at how they work together to generate memory and other brain functions.

THE ATOMS OF THE BRAIN

The brain consists of a large number of active cells, called neurons (making up the gray matter), which process and produce electrical and chemical currents. There are on the order of one hundred billion (10^{11}) to one thousand billion, or a trillion (10^{12}), neurons in the human brain. The brain has many other cells, which nourish and (physically) support the neurons, but the neurons are the most interesting cells. They are thought to be responsible for cognitive and memory functions. The neurons are intricately connected to each other by fibers, which make up the white matter of the brain. These fibers allow neurons to communicate with each other by electrical signals.

Each neuron consists of a body (called the soma) and tentacles (called dendrites), which act like tree roots, seeking and receiving information from thousands of other neurons. Information is collected by the dendrites (and the soma itself) and transmitted to the soma, where it is processed. If the sum of all of these inputs into any particular neuron is of sufficient magnitude (greater than some inbuilt electrical threshold) and of sufficient synchronicity (that is, the signals arrive at almost the same time), that neuron will itself "fire." It will then create an electrical pulse, which will be transmitted down its outlet (called the axon) to other neurons. Typically an axon is about one millionth of a meter in thickness when it leaves a neuron. Farther away from the neuron, the axon branches out, and this enables the signal from the excited neuron to be communicated to thousands of other neurons. Some neurons can even transmit their signal to around one million other neurons, and up to a meter away (such as from sensory neurons in the feet to neurons in the spinal cord).

There are many different types of neurons, including pyramidal neurons, which are large excitatory neurons whose axons can span distances on the scale of the brain; Purkinje neurons, which are found in the cerebellum (a particular structure inside the brain concerned with movement) and have an extensive dendritic treelike structure; and stellate neurons, which have a fanlike branching axon structure. Neurons are also classified as to whether they are sensory neurons, motor neurons, or interneurons. Sensory neurons convert physical input from the environment—such as light, sound, temperature, and pressure—into electrical signals. (It is quite amazing how nature has developed such sophisticated miniature devices that convert various forms of energy from one state into another.) The electrical signals generated by the sensory neurons are transferred to the brain, where they are processed by the interneurons. This processing initially involves extracting certain features about the input data, followed by the "comparison" of this processed information with other stored memory. There is actually no direct comparison with memory taking place; what is happening is that the information is being processed according to previous experience. (See the discussion in chapter 3.) Interneurons send signals backward and forward to each other until they reach a decision, which is relayed to the motor neurons, which in turn convert the electrical signals into kinetic energy in the form of muscle contractions. Sometimes a decision made by the interneurons is just stored away in the brain and does not result in any motor action.

The Spanish neuroanatomist Santiago Ramón y Cajal (1852–1934) was the first person to identify neurons and to study the way in which they are intricately connected to each other, back in 1908. Cajal produced painstaking drawings of neurons and neural network architectures in the human brain using a method called Golgi staining, where a selection of neurons are stained with ink. Cajal put forward the neuron doctrine, maintaining that the nervous system is made up of neurons, each of which could be considered a separate independent entity consisting of dendrites, a soma, an axon, and axon branches. He suggested that these neurons somehow communicate with each other where axons touch the dendrites of other neurons. He further posited that the flow of information (or electricity) is from the dendrites to the cell body, down the axon, and on to the next neuron.

NEURAL NETWORKS

Figure 2.1 shows a schematic diagram of a neuron with only a few inputs and outputs, many fewer than in a real biological system. It also shows various electrical pulses moving around the neurons at a particular moment in time. These electrical signals move from a dendrite to the soma, and if the neuron fires, an electric signal moves down its axon to other neurons. The brain is an extremely complicated network of neurons, interacting with each other simultaneously. It is arguably the most complicated system known to man.

To illustrate the complexity of neural network systems, or networks of interconnected neurons, figure 2.2 shows a very simple neural network with only nine neurons. Here, each neuron communicates to only a handful of other neurons, some belonging to this set and others from outside the set. For example,

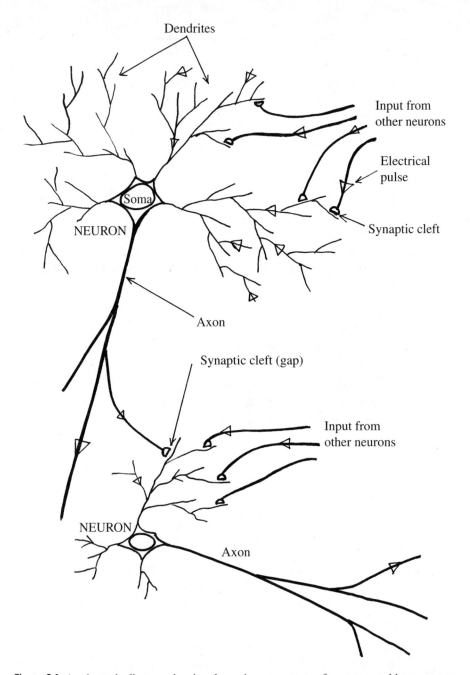

Figure 2.1. A schematic diagram showing the main components of a neuron and how neurons communicate with each other. Electrical signals are transferred between neurons at the synaptic connections. An electrical signal travels up a dendrite of a receiving neuron toward its soma (main body). The electrical signals from a number of neurons are delivered to the soma, where they are assessed, and if this neuron receives sufficient excitation, it will "fire" and transmit its own electrical signal to other neurons down its axon (outlet). The arrows in the diagram indicate possible positions of electrical pulses traveling through the network shown, at a particular moment in time.

neuron 2 sends a signal to neurons 4, 5, 7, and 9 and to two other neurons outside the network, and receives input from neurons 1, 3, and 7. One can imagine how complicated this can get in a real biological system, where each neuron communicates to thousands, or even tens of thousands, of other neurons. In general, the direction of flow of neural charge is not symmetric, as is illustrated in figure 2.2. Neuron 2 sends a signal to neuron 5, but neuron 5 does not send a signal directly back to neuron 2. The brain is a very intricately connected system of neurons. Figures 2.3 and 2.4 show photographs of real neural networks taken with a scanning electron microscope. They clearly demonstrate the complex way neurons are connected with each other.

The situation is further complicated by the fact that the signals arriving at neurons from other neurons are not of equal strength or intensity. This is a result of the way that memory is stored in a neural network. The amount of electric charge (or more appropriately the electrical depolarization) that is transmitted from one neuron to another depends on the strength of the connection between these neurons. What is more, this connection strength can be varied with experience, as we shall see.

Another complication with real neural networks is that most neurons receive inputs from other neurons at different times. Neurons have a certain "threshold potential" that must be exceeded before they fire. A typical neuron will fire if it receives excitatory input from a hundred or so other neurons in a relatively short time period, on the order of a millisecond or so. After a neuron

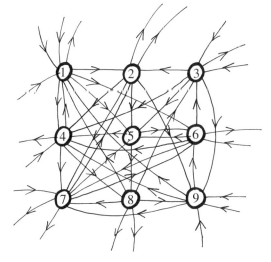

Figure 2.2. A simple neural network with only nine neurons. Here each neuron connects to a handful of other neurons, whereas in a real biological system each neuron connects to thousands or tens of thousands of other neurons. The arrows in this diagram show the possible directions of communication. In real neural networks, communication between neurons is usually unidirectional.

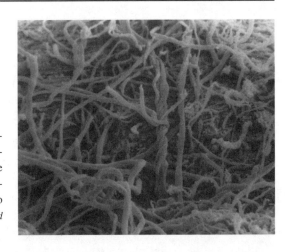

Figure 2.3. A scanning electron microscope picture showing the elaborate entanglement of axons (white matter) in the cerebral cortex of human tissue. Magnification: 8,000 to 10,000. *(Picture courtesy of Dr. Arnold Scheibel, UCLA.)*

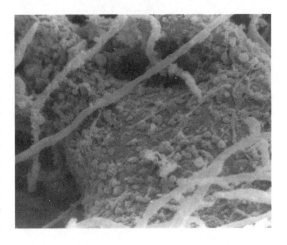

Figure 2.4. A scanning electron microscope picture showing an array of synaptic boutons (axonal endings from other neurons) on the surface of the soma (main body) of a single pyramidal neuron. The main (apical) dendrite of this pyramidal neuron extends obliquely toward the top left-hand corner of the picture. Magnification: 8,000 to 10,000. *(Picture courtesy of Dr. Arnold Scheibel, UCLA.)*

fires, it takes about 3 or 4 milliseconds for the neuron to readjust its electrical potential so that it can begin to process new input again. During this "refractory" or rest period, a neuron is unable to fire. The fact that a neuron may sometimes not fire, such as when it does not receive sufficient excitation, is what actually gives the brain its computational abilities. A system that reacts linearly, or predictably, is not capable of nontrivial computation.

There are also two different ways in which a neuron can act on other neurons. It can excite other neurons when it gets excited. It is then called an excitatory neuron. Alternatively, a neuron can inhibit other neurons when it gets excited. It is then called an inhibitory neuron. Inhibitory neurons are especially useful in regulating motor control and processes like habituation, where organisms cease to respond to repeated exposure to a benign stimulus. In general, a neuron acts in one way or the other; that is, it either excites all neurons it sends a signal to, or it inhibits them all. This is known as Dale's law. Note however that a neuron can receive both excitatory and inhibitory input from other neurons.

THE SYNAPTIC GAP

Neurons communicate with each other not only by the transfer of electricity, but by the transfer of chemicals as well. These chemicals are transferred at the synaptic cleft (or the synaptic gap), where two neurons actually connect to each other (see figures 2.1 and 2.5). The amount of chemical current transferred across the synaptic cleft directly influences the amount of electricity transferred to the other neuron. After an axon leaves a neuron, it branches out in thousands of directions. Each of these tentacles ends in a small "bouton," which connects itself to a dendrite or soma of another neuron. (Some axons synapse onto the axons of other neurons, forming what are called heterosynaptic connections. Typically these connections have no effect on the electrodynamics at the soma of the cell to which they are connected, although they can indirectly influence the amount of neurotransmitter that this cell releases to other neurons in the vicinity of the connection. In most models these types of secondary connections are ignored.)

The synaptic gap between neurons is typically about 0.02 millionths of a meter wide; put another way, if we put five thousand synaptic clefts end-on-end, this would be equivalent to the thickness of a sheet of paper (0.1 millimeter thick). At this junction between two neurons, the electrical pulse from the input neuron is transformed into a chemical current, which consists of small organic molecules called neurotransmitters. (See figure 2.8 below for some of the main neurotransmitters in the brain.) The electrical stimulation causes neurotransmitters to be released from tiny spherical structures (called vesicles) located on the transmitting neuron (also called the presynaptic cell). When a vesicle is opened, it usually releases all of its contents into the synaptic cleft. It is thought that each vesicle may hold around five thousand neurotransmitter molecules (Katz 1969; Squire and Kandel 1999). The released neurotransmitter drifts across the synaptic gap to the receiving neuron (also called the postsynaptic cell), where it is detected or greeted by so-called neuroreceptors. The

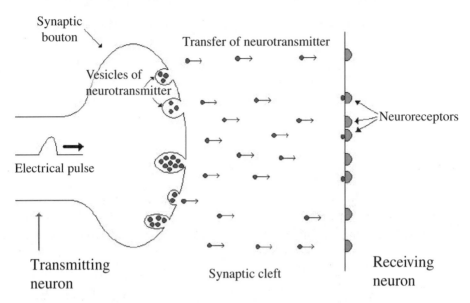

Figure 2.5. Neurons communicate with each other via the synaptic cleft. An electrical spike arrives at the synaptic bouton of the presynaptic neuron, where it stimulates the release of neurotransmitter from tiny vesicles. This neurotransmitter drifts across the synaptic cleft to the postsynaptic neuron, where it is greeted by neuroreceptors, which transform the chemical signal back into an electrical signal. The electrical signal then propagates to the soma of the postsynaptic cell, where it is processed.

neuroreceptors are tailor-made to receive the specific neurotransmitters that make the journey across the synaptic cleft. This chemical current is then transformed back into an electrical current, using secondary messengers. The electrical signal travels down the dendritic branch to the soma of the receiving neuron. This process is illustrated in figure 2.5. The way that the soma or body of a neuron works is yet another story in itself, which involves the transfer of sodium and potassium ions across cell membranes.

The amount of chemical current that is actually transferred or released at the synaptic junction depends on how often that particular junction has been used in the past. This is the essence of how memory is stored in a neural network system. If a neuron is repeatedly excited, it releases chemicals (like nitric oxide, NO), which drift back to the presynaptic neurons that were exciting it, and this causes those neurons to increase the available amount of neurotransmitter for future transmission. This will ensure that the next time a signal travels down one of these channels its effect will be amplified, and in that way the same pattern of neural excitation that caused this change will be recalled. This is the essence of how memory is stored in the brain. If there were no synaptic clefts or adjustable chemical currents, it would not be possible to store memory. It has been estimated that there are about one hundred trillion (10^{14}) to one thousand trillion, or a million billion (10^{15}), synaptic connections in the human brain.

NEUROANATOMY

The connectivity of neurons in the brain is largely random, but the brain does appear to have some structure. First of all, the brain seems to be composed of a number of organs. Figure 2.6 shows some of the main neuroanatomical structures, viewed from the inside of a bisected human brain. The brain divides into two almost identical halves, so there are actually two of every brain organ

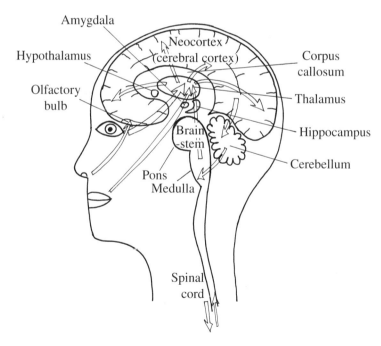

Figure 2.6. Some of the most prominent neuroanatomical structures in the human brain, also showing the general direction of sensory information flow.

(except for the pineal gland), one located in the left hemisphere and one in the right hemisphere. It is not understood why this is so. Perhaps each side serves as a backup for the other, or maybe the two halves are required for some sort of stereo sensory three-dimensional processing. The other important thing about the two hemispheres is that the left hemisphere processes information from, and the controls the movement of, the right side of the body. Similarly the left side of the body is directly connected to the right cerebral hemisphere. Incidentally, the seventeenth-century French philosopher René Descartes thought that the unique pineal gland, which is located almost in the middle of the brain, might be intimately involved with the mind or the soul. This is not the case, although it is still not clear what the function of the pineal gland is other than the production of the mysterious hormone melatonin, which is implicated with sleep.

Each of the dual anatomical structures in the brain is thought to serve an important role in relation to brain function. The cerebral cortex, or the neocortex, as it is also called, is the outermost layer of the brain. It consists of a folded sheet of gray matter about 2–3 millimeters thick, with an area of about one square foot unfolded. The neocortex is thought to be the center of our most impressive mental functions, such as memory, cognition, language, and consciousness. It is quite remarkable that these extraordinary brain functions should reside in such a small, thin slab of neural tissue. The cerebral cortex in humans, compared with the other brain organs, is much larger than in other mammals that have a similar brain structure. The other parts of our brains are comparatively similar to those of other mammals, and most of these other primitive structures are present in many vertebrates. This is why the cerebral cortex in humans is also referred to as the neocortex (which means "new cortex"): because in an evolutionary sense it is new.

All sensory input, such as from the eyes, ears, and body (including sensations of heat, pressure, and pain), passes through the thalamus en route to the neocortex, where it is processed. The thalamus is located centrally in the brain, just underneath the neocortex, and acts as the gateway to the neocortex. Each type of sensory input passes through a specific part of the thalamus before it is transferred to specific areas in the neocortex.

The hypothalamus, which is located just under the thalamus, is thought to be associated with controlling body temperature, hunger, thirst, sexual activity, and the release of certain hormones (relating to growth, reproduction, and various metabolic activities) from the pituitary gland into the bloodstream. The hypothalamus is also implicated with emotions like fear.

The hippocampus is a small seahorse-shaped body located under the thalamus, which is involved in converting new short-term memory into medium or long-term memory and acts as an addressing system to recent memory. A famous patient called H. M., who had his hippocampus bilaterally removed (meaning on both sides), was unable to remember details of what transpired just minutes earlier and was consequently unable to lay down long-term memory. (See the discussion in chapter 3 for more details.) The hippocampus is also interesting because it is one of the main targets of internal stimulation during dream sleep. (See the discussion in chapter 5.)

The amygdala is a small almond-shaped region in the brain that is highly implicated with emotion, fear, and emotional memory. It has connections to the hypothalamus and so can influence the release of hormones from the pituitary gland, which can affect our emotional states. If the amygdala is removed from a monkey's brain, the monkey seems to become fearless. Patients with lesions to the amygdala seem to remember things, but without the emotional content of the memory. Like the hippocampus, the amygdala is thought to act as an addressing system for emotional memory in the cortex.

The cerebellum is a cauliflower-shaped region at the base of the brain that is thought to be concerned with movement and balance, and the learning of new skills, and motor actions.

The brain stem is the extension of the spinal cord into the brain. Among other things, the brain stem is the gateway from the body to the brain and from the brain back to the body. It brings together all sensory information from the body that enters the brain, and distributes the brain's instructions, such as motor actions, back to the body. The brain stem is also the control center of our life-support systems, such as the respiratory and cardiovascular systems. Damage to the brain stem usually results in death, whereas damage to other areas in the brain, such as the cerebral cortex, may just result in impaired memory. This may be one of the reasons the brain stem is located deep inside the brain. The brain stem is also thought to play a prominent role in other natural rhythms of the body, such as the sleep/wake and sleep/dream cycle. In addition, the brain stem has neurons that regulate the intensity of our learning experiences. This is achieved by the release of neurotransmitters into the neocortex when something is to be learned. The group of a few thousand neurons that is responsible for this is called the "locus coeruleus." Each neuron in the locus coeruleus can connect to up to one million neurons in the neocortex, and so can directly influence thousands of millions of neurons there. This group of neurons is also thought to maintain arousal, and is implicated with our waking consciousness.

Parts of the brain stem also act as the generator of our dreams. (See chapter 5 for more details.) A small group of neurons in the brain stem, called the "REM-on" cells, located in the top part of the brain stem called the pons (meaning bridge), become very excited during rapid-eye-movement (REM) sleep, the phase of sleep associated with dreaming. The apparently random signals generated by these neurons bombard the neocortex and other structures. Allan Hobson and Robert McCarley have suggested that dreams are a result of the brain, specifically the neocortex, trying to make sense of the chaotic input generated by these brain-stem neurons (Hobson and McCarley 1977; Hobson 1988). The brain stem has another small group of neurons that stop us from acting out our dreams. These neurons prevent motor actions generated in the brain during REM sleep from being sent down the spinal cord. If they are surgically removed from a cat, the cat is seen to move about and act out its dreams while it is asleep and dreaming. (See chapter 5 for more details.)

On closer examination, just as the brain stem has a number of different regions that perform special functions, all of the anatomical structures mentioned previously have substructures of their own. We have already noted that the thalamus has groups of small populations of neurons (called "nuclei") that handle specific forms of sensory input. One of the nuclei in the thalamus handles visual input, another handles auditory input, and another handles somatosensory information relating to touch and pain. The thalamus also seems to be involved in attention and consciousness. One of the main reasons for this is that there are reciprocal connections from areas in the neocortex back to the thalamic nuclei, which relay signals to these specific cortical areas. It is as if each of the nuclei in the thalamus needs to know what the corresponding bits of the cerebral cortex are doing.

Parts of the hypothalamus perform the special functions mentioned earlier. In addition the hypothalamus is involved in the resynchronization of our sleep/wake cycle to twenty-four hours. When humans are placed in underground chambers isolated from the outside world with no sunlight, their natural daily rhythm, called the circadian rhythm, is about twenty-five to twenty-six hours. It is thought that a small group of neurons in the hypothalamus called the suprachiasmatic nucleus helps to achieve the resynchronization of our clocks back to twenty-four hours by using cells that are hypersensitive to the presence of sunlight in the visual field. This nucleus is located just above the place (called the optical chiasm) where visual information from both eyes comes together before it is distributed to the neocortex. The information gathered by the suprachiasmatic nucleus is relayed to the brain stem, where natural body rhythms are controlled. Our circadian rhythm is usually upset when we travel long distances in airplanes, leading to the condition called jet lag, in which we feel lethargic and tired. Depending on the time of arrival, one of the best ways to get over jet lag may be to go outside and walk around in the bright sunlight, instead of sleeping it off during the day.

The neocortex is itself organized into regions that process specific types of input or specialize in different attributes of cognitive function, such as reading, speaking, and thinking (See figure 2.7). For example, areas on the parietal lobes are involved in the sense of touch (somatosensory input), the occipital lobes are involved in vision, the frontal lobes are involved in personality and thinking, and the temporal lobes are involved in general memory. Other more

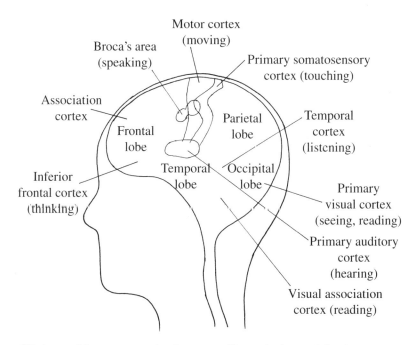

Figure 2.7. Areas of the neocortex, showing some of its particular specializations.

specific areas within each of these regions are involved in even more specific brain functions. A small area in the frontal lobes, for example, called Broca's area is known to be involved in speech. Most of our knowledge about what different parts of the human brain do comes from the examination of people who have had some part of their brain surgically removed, either because of an accident or a brain tumor, or to control epilepsy (excessive electrical activity in the brain). Scientists realized that our personality and our ability to plan are housed in the frontal lobes, because people who had a large part of their frontal lobes removed, such as in prefrontal lobotomies, literally became zombies after the operation, devoid of any personality.

If one looks even closer at the neocortex, it is organized into layers, columns, and other structures. The neocortex is arranged into some six or more layers, which are intricately connected to each other, mainly by short-range connections. Each of these layers is thought to be associated with various levels of information processing. The level of processing does not simply proceed in one direction, but jumps around from one layer to another. In infants, however, the innermost layers are developed first. Most sideways connections in the cerebral cortex are short-range and inhibitory, while most of the information that is carried far away, to other cortical areas or outside the cerebral cortex, is transmitted by long-range excitatory pyramidal neurons.

In 1957, Vernon Mountcastle discovered that the cerebral cortex—in particular the somatosensory cortex, which is associated with touch and bodily sensations—is organized into columns, which seem to have specific functions (Mountcastle 1957). Other parts of the cerebral cortex are also organized in this way. The visual cortex, for example, has columns that specialize in eye dominance and the detection of horizontal and vertical lines. There are lateral connections among these various columns, but as noted above, most of these connections are short ranging.

We are not particularly concerned here with the fine details of the way the neocortex is organized, except to note that it (and other parts of the brain for that matter) seems to consist of many small neural networks. Each of these subnetworks, or modules, contains on the order of ten thousand neurons. The modules are themselves organized into larger groups of networks and ultimately into brain organs.

Finally, the corpus callosum is an enormous bundle of white fibers that connect the left and right hemispheres together. It is thought to contain around 200 million axons, which means that, as each axon can branch into ten thousand or so outlets, it traffics information for up to 2 trillion synaptic connections. The corpus callosum has been severed in some patients who suffered from epilepsy, to try to control the level of electricity in their brains. These patients with two separate brains are at times observed to act as if they have two different personalties, or two separate minds, although the dominant side of the brain (usually the left), which is the one that houses language, wins out.

THE CHEMICAL BRAIN

The brain has a number of different neurotransmitter or chemical systems, each implicated with different attributes of brain function. The main neurotransmitters in the brain are gamma-aminobutyric acid (GABA), glutamate, norepinephrine, acetylcholine, dopamine, and serotonin. Neurotransmitters are relatively small organic molecules, whose chemical structures are shown in figure 2.8. Individual neurons generally use only one neurotransmitter, so they are also labeled according to their chemical nature.

Dale's law, that neurons act in either an all excitatory or all inhibitory way, is thought to be based on the fact that a neuron operates with one neurotransmitter at all of its presynaptic terminals, where it communicates its signal to other neurons (Eccles 1964). This notion, however, seems to be contradicted by the fact that neurons can accept signals from neurons utilizing different neurotransmitters. This means that neurons must also have the appropriate neuroreceptor systems to interpret these chemical signals.

GABA is an inhibitory neurotransmitter and is the most common neurotransmitter in the brain, with about one third of all neurons (or more appropriately, synapses) using this chemical messenger. People who suffer from Huntington's chorea have a shortfall of GABA, and one of the main symptoms of this disease is uncontrollable involuntary movements, resulting from the lack of adequate inhibitory control.

Norepinephrine is the main neurotransmitter of the locus coeruleus. It is an inhibitory neurotransmitter that is distinctly implicated with learning and arousal. When we are awake and learning, norepinephrine is at its maximum level of activity, and when we are asleep and dreaming, norepinephrine (or more appropriately, the neurons in the locus coeruleus that use this neurotransmitter) is essentially absent (or dormant). This is consistent with the view that the usual learning processes are not operative during dream sleep, an interesting point because it has been suggested that the brain may be engaged with an unlearning process (or the opposite of conventional learning) during this phase of sleep (see chapter 5).

Acetylcholine was the first neurotransmitter to be discovered by Otto Loewi in 1921. It is an excitatory neurotransmitter, which is implicated with memory and is used in peripheral nerves to interact with muscles. Most of the "cholinergic" neurons (meaning neurons that use acetylcholine as their messenger neurotransmitter) are located in the basal nucleus (or nucleus basalis of Meynert), which is positioned at the base of the brain, close to the amygdala. The REM-on cells in the brain stem, which become excited when we dream, also utilize acetylcholine. People who suffer from Alzheimer's disease have an acute shortage of cholinergic neurons in the parts of the brain where most reside: the basal nucleus, the limbic system (which includes the hippocampus and the amygdala), and some parts of the cerebral cortex. Alzheimer's disease is a degenerative disease that affects the elderly, initially involving a severe deterioration of short-term memory, followed later by other forms of memory. As we saw earlier, the hippocampus and the amygdala, in the limbic system, are important for the medium- and long-term storage of memory. The memory loss seen in Alzheimer's disease sufferers is similar to the memory loss seen in the patient H. M., who is without a hippocampus.

Serotonin is another inhibitory neurotransmitter, whose center is located in the raphe nuclei, in the brain stem. Serotonin and norepinephrine are collectively referred to as "aminergic" neurotransmitters because of their similarity to each other and to chemicals called biogenic amines. They are both thought to be strongly linked to learning. When the levels of serotonin fall, such as during dream sleep or when one is under the influence of psychedelic drugs (like LSD [lysergic acid diethylamide]), the brain runs with uncontrolled imagination and fantasy. Low levels of serotonin are also implicated with depression. Prozac (fluoxetine hydrochloride), which acts as a serotonin re-uptake inhibitor and retards the process of reabsorbing serotonin back into the presynaptic vesicles, is used to treat depression. This is interesting because depression is also linked with excessive dreaming, which is itself associated with low levels of serotonin.

Glutamate, or glutamic acid, is the main excitatory neurotransmitter in the cerebral cortex. No drugs have been found that affect glutamate.

Dopamine is mainly concerned with the control of motor function. People with Parkinson's disease, which is normally associated with tremors and the inability to move, are usually found to have a shortage of dopamine in the corpus striatum, a region in the brain involved in the regulation of movement. Excess levels of dopamine are implicated with schizophrenia, or madness. Dopamine is also thought to be linked with the brain's pleasure and reward systems.

Histamine, which is involved in allergic reactions in the body, is also a neurotransmitter that is thought to be related to emotional behavior. Some antihistamines have been used to sedate schizophrenic patients, without putting them to sleep.

Glycine is the most prominent neurotransmitter in the spinal cord and the brain stem. It is an inhibitory neurotransmitter.

There are many other neurotransmitters, probably around thirty or more. There are also neuroreceptors, which detect the amount of neurotransmitter that makes it across the synaptic cleft, and secondary messengers, which transform a neuroreceptor's message into an electrical signal for the postsynaptic neuron. In addition the brain uses a range of other chemicals to manufacture, break down, and transport neurotransmitters, and chemicals that travel from a postsynaptic cell to a presynaptic cell to tell it to produce or make available more neurotransmitter. The brain is a complicated "soup" of chemicals, which may also interact with each other in subtle ways.

Drugs can interfere with brain function, either directly or indirectly, by influencing the amount of neurotransmitter that is released, transferred across the synaptic cleft, or detected by the neuroreceptors. For example, drugs can block the release of neurotransmitters from the vesicles, or they can cause the vesicles to prematurely release their neurotransmitters; they can mimic the neurotransmitters and block the neuroreceptors from detecting the real neurotransmitters; or they can upset the process by which the chemical signal is transformed into an electrical signal in the postsynaptic cell. The action of drugs is further complicated by the fact that they can act on more than one neurotransmitter system. Drugs can also act on secondary brain chemicals, which are used to produce, destroy, or transport neurotransmitters back to the presynaptic cell, or they can interfere with the learning process itself. They consequently affect memory and other aspects of brain function like emotion, mood, motor control, and even consciousness. Drugs are interesting because they provide another window into understanding how the brain works (Snyder 1986).

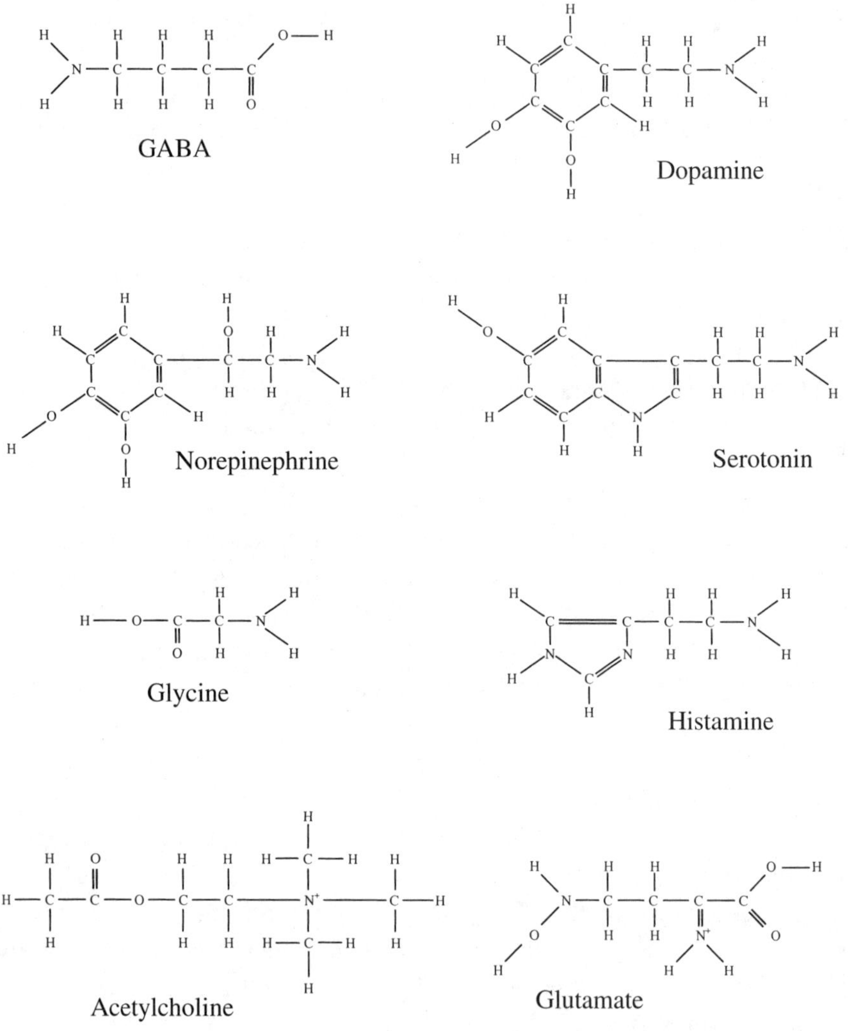

C = Carbon
N = Nitrogen
O = Oxygen
H = Hydrogen

Figure 2.8. The chemical structure of some of the main neurotransmitters in the brain.

THE DEVELOPING BRAIN

Neurons are much like other cells in the body. They divide and produce new cells, at least in the womb. The fetus gains neurons at the rate of thousands every second, and a newborn has practically all of the one hundred billion neurons that it has in adulthood. An infant probably has even more neurons than an adult, because after birth neurons essentially cease to divide and reproduce new neurons, and some neurons die. Studies of brain tissue of infants who have died prematurely show that the key difference between the brain of an infant and the brain of an adult is that a newborn has only a fraction of the synaptic connections that are present in an adult brain. These connections are where memory and life experiences are stored.

Although most of the connections in the brain seem to be random, some, such as the way information flows from the eyes to the visual cortex via the thalamus, are quite precise. Our genes probably direct the broad organization of the brain, such as the main connections and the division of the brain into organs, modules, and nuclei. Most of the important connections are either present at birth or take place during the early stages of the development of the brain immediately after birth. With these basic structures in place, the rest of the wiring in the human brain, particularly in the neocortex, is made in an almost random manner, with some fine-tuning and pruning as we interact with the environment around us. This is thought to be the case because there seems to be insufficient information in DNA (only about thirty thousand active genes) to tell the brain how to make hundreds to thousands of trillions of connections (Changeux 1986). (One needs to be a little careful making this statement, because although, for example, the digits of the mathematical number π are infinite and appear to be random, one can write a program, with only a few hundred bits of information, that can generate this string of digits [John Hopfield, private communication]. Another example is afforded by chaos theory, where simple mathematical equations (like the logistic map, or Mandelbrot's algorithm) can generate an infinite complexity. Such programs may also be contained in DNA.)

The rapid growth of synaptic connections in the infant brain can be inferred from the rapid increase in brain weight with age, since most of the increase in brain weight can be attributed to the growth in axons and dendrites by the immature neurons. The weight of a newborn's brain is approximately

330 grams, and this increases to about 1,000 grams by the age of two years, followed by a much slower rate of increase until the age of fourteen years, when it is thought that the brain has reached maturity (approximately 1,400 grams). If one assumes that the newborn's brain has practically no synaptic connections, which although unrealistic is not a bad assumption for the purposes of this calculation, and that by fourteen years of age it has one million billion connections, one can conclude that one gram of brain weight corresponds to approximately a thousand billion synaptic connections. This means that in the first two years of life, which correspond to approximately 63 million seconds, an infant makes approximately 670,000 billion synaptic connections. This amounts to making a staggering 10 million connections every second for every moment of the day, including the times when the infant is asleep. It has been suggested that the human brain makes even more connections than this, with many of these new synaptic connections pruned along the way as the brain self-organizes to optimize its interaction capabilities with the environment. From the age of two years to fourteen years the growth in synaptic connections slows to about one million connections per second, which is itself still considerable. Although it is known that adults lose neurons with age, there is no reason why some synaptic growth does not also persist, although one would expect that this would be at a much lower rate than that seen in infants and children.

There are good reasons why most of the synaptic connections are not made while the baby is in the womb. First of all, there is insufficient time to make all of these connections, and most of the neurons have not even been created as yet. More importantly, though, most of the synaptic connections, which themselves are where memory will be stored, need to be made while we are interacting with the environment and other people. If the human brain were completely connected at birth, it would be far less capable of adapting to unforeseen circumstances and new challenges to survival. Finally, as we have seen, there appears to be insufficient information in DNA to tell the brain how to hard-wire itself. The majority of the brain's connections must be made randomly and in accord with experience.

Although most synaptic connections may be random, many, particularly the important precise connections as discussed earlier, are guided somehow. Genes are thought to play a prominent role in making these connections, which are required to develop a basic brain that can compute and process information. Ramón y Cajal observed that most neurons sprout dendrites and axons in tree-like structures, in certain well-defined directions, or cones as it were, as if they "knew" where they should go. It has been suggested that these growth cones may be influenced by the release of certain chemicals in the brain emanating from regions to which the dendrites and axons should grow, but it is difficult to explain long-range connections in this way. In some cases these connections can be quite precise, whereas in most other cases the connections are not particularly specific.

At one time it was thought that the entire brain might be genetically connected. The Nobel laureate Roger Sperry severed the optic nerve in fish and frogs and found that the thousands of individual nerve fibers reestablished the same connections that existed before the operation. This suggested that each fiber acted as if it were chemically different from every other fiber, as the identical fibers were able to find each other again. After these fibers had regrown, the animal was able to see as if nothing had happened. This led Sperry to conclude that synaptic connections must be controlled by genes, but current thinking has it that most of the hundreds of trillions of connections in the brain are made randomly. Although our genes undoubtedly determine certain important connections, such as where the optic nerves attach, most connections are made, and must be made, randomly or semirandomly. The brain perfects its architecture by pruning unnecessary circuits.

Another good reason why the brain may not be hard-wired at birth is that if it were, it would be more vulnerable when an incorrect connection was made. If we were to reassemble a transistor radio and put the components in the wrong places, it would not work. The random self-organizing nature of most of the brain's connections gives it added flexibility and insurance against such problems.

Jean-Pierre Changeux has proposed that the infant brain actually makes even more synaptic connections than the estimates given above. According to Changeux, the infant's brain makes many semirandom connections, and as the infant learns from the environment, some of these new synaptic connections are strengthened, while others, which are used less or are not required, become redundant and are pruned (Changeux 1986; Changeux, Courrège, and Danchin 1973). In addition to pruning redundant synaptic connections, the brain may also dispose of some redundant neurons. This would free up more space for the growth of new axonal branches and dendrites. The electron microscope photos in figures 2.3 and 2.4 show how intricate the connectivity is in real neural networks. Changeux suggests that spontaneous activity in the neural networks determines which connections, made during the period of transient redundancy,

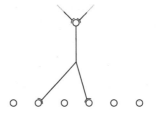

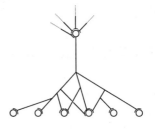
Period of transient growth

Figure 2.9. Changeux's theory of "epigenesis" is that after the basic neural circuits of the brain have been developed, the brain is driven by periods of exuberant semirandom growth in new synaptic connections leading to transient redundancy. These periods of rapid growth are followed by periods of selective stabilization and regression through the pruning of redundant connections. In this process some redundant neurons may also die off.

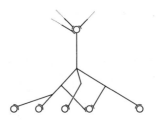
Pruning redundant connections

can be pruned. Connections that are not used very often or are not required may be the first to go. This trend of synaptic overgrowth followed by subsequent pruning, as illustrated in figure 2.9, is posited to occur in numerous waves from birth to puberty. How the brain organizes itself and knows which circuits to keep and which to discard is not yet understood.

We should note that the fetus could also utilize the paired process of random transient growth of synaptic connections followed by subsequent pruning of redundant connections. A fetal brain could make numerous semirandom semispecific connections that could be tested, refined, and pruned in some way to best express the wishes or information housed in its genes. It is reasonable to think that something like this is taking place, because it seems unlikely that the brain has sufficient chemical resources to direct the precise connectivity of so many important parts of itself.

THE CRITICAL PERIOD

As noted previously, the neocortex is organized in columns that seem to perform specialized functions. In the visual cortex, for example, there are columns that exclusively detect vertical and horizontal lines or edges. (Visual information from either eye is processed in both the left and right hemispheres. Information from the left visual field [on your left side], which can be collected by either eye, is processed in the right hemisphere, and visual information from the right visual field is processed in the left hemisphere.) It is known by examining the brains of very young kittens (before they have opened their eyes) and comparing these columns with those in the brains of older kittens that most of the circuitry and organization of these cortical columns are already developed at birth. This suggests that genes play a prominent role in this organization. However, there are a number of experiments suggesting that experience, especially soon after birth and within a certain period (called the "critical period"), which is different for different animals and maybe different for different senses, can modify some of these important connections and change the functionality of these specialized cortical columns.

The critical period in cats is on the order of twelve weeks. The Nobel laureates David Hubel and Torsten Wiesel found that if a kitten has one eye shut for a period extending over much of this critical period, the columns in the visual cortex normally associated with the closed eye are not properly developed (Hubel 1988; Hubel and Wiesel 1970; Wiesel 1982; Winson 1985). The damage to the kitten's visual system associated with the closed eye is irreversible, and the kitten becomes blind in that eye. If an adult cat undergoes a similar experiment, this does not happen. Normal vision is restored soon after the eye is reopened. This blindness observed in kittens is a result of damage not to the kitten's eye, but to the kitten's visual system in the brain. This suggests that during the critical period in the early development of the kitten's brain, some of the neural circuits are being hard-wired with the help of input from the environment. These important connections are not chemically driven but are refined through experience. What also happens is that the other eye, which can still see, or process visual information, utilizes those neural structures that were originally reserved for the shut eye. This suggests that there is some competition at play in the development of the central nervous system, particularly in the early stages of development when some circuits are still being hard-wired.

In other experiments (on poor little kittens again), Colin Blakemore and Grahame Cooper placed kittens in environments in which they were exposed solely to either vertical or horizontal lines (Blakemore and Cooper 1970; Winson 1985). They found that if a kitten was exposed only to horizontal edges during most of the critical period, then after the period had expired and the kitten was placed in a normal environment, it could not detect vertical edges. If a horizontal bar was placed in front of the kitten, it would avoid the bar, but if a vertical bar was placed in front of the kitten, it would walk straight into the bar, as if it had not seen the bar at all. The same phenomenon was observed when a kitten was exposed exclusively to vertical lines during the critical period. In this case the kitten would be able to see vertical bars placed in front of it, but would be unable to see a horizontal bar. Once again, if the neurons that are normally reserved to a particular type of line orientation are not exposed to experience during the critical period, they are utilized by the other specialized brain functions that the brain needs to develop during this critical period.

In humans the critical period is on the order of two years, which is the period of most rapid synaptic growth in the human brain. If a child had one of its eyes covered for most of this period, it would be blind in the covered eye. Newborns with cataracts who are not treated within the critical period remain blind. On the other hand, adults with blindness due to cataracts (for up three years) have their vision restored after an operation (Hubel 1988).

The critical period also applies to the other senses. For example, if during the first two years of life one is not exposed to the phonetic sounds associated with the Chinese language, then one will not be capable of properly developing these sounds later. Nor will one be able to distinguish the subtle differences in Chinese phonetics. The same applies to other languages such as Russian, and the reverse is also true, so that if a Chinese person is not exposed to certain English sounds during the critical period, that person will have difficulty in correctly making those sounds later. The critical period varies for different forms of sensory information.

The human brain seems to be quite malleable during childhood, beyond the critical period. This may be linked with the period of slower synaptic growth between the ages of two and fourteen. It is known, for example, that if young children are trained in certain activities at an early age, they may become very skilled in those particular activities. Children in the former Soviet-bloc countries, like Russia and Romania, were often exposed to gymnastics and classical music at a very early age, in the hope that they might become elite athletes or talented musicians. It is also common knowledge that what we do as we grow up determines our career to a large extent. Finally, children are much more capable of learning languages than adults.

LEARNING AND FLEXIBILITY

In the next chapter, I will show how learning can take place in neural networks through the adjustment of the strengths of the synaptic connections. This plasticity is realized by controlling the amount of neurotransmitter that crosses and is detected on the other side of the synaptic cleft. Changeux's theory suggests that learning may also take place as the brain goes through phases of exuberant synaptic growth and pruning. This process can be likened to Darwinian selection, with the fittest and most active connections winning out over others.

Synaptic pruning presumably allows an immature infant to learn at a much faster rate than an adult. This is all very well, because an infant needs to learn so much more during its early years. It needs to learn how to walk, grasp with its hands, speak, respond, understand language, and see. The added flexibility in its development provided by periods of synaptic growth and pruning may also enable the infant to better cope with new challenges, or changes in environment that its parents and ancestors may not have had to deal with.

The added flexibility of the neural system afforded by random connections may also partly explain why we are able to perform certain acts that do not have an obvious biological function. An example of this is our ability to learn how to skateboard with exceptional skill and balance. There are no obvious biological reasons why we should have the predetermined neural capabilities to perform such an action so well. This general phenomenon troubled Alfred Wallace, one of the codiscoverers of natural selection (Cziko 1995). Wallace reasoned that we should not have any special abilities that could not have been useful to our ancestors. He was extremely puzzled, for example, that Africans could so easily learn to play Western or European music. There was nothing in their natural environment or history that would encourage their ability to do this. Natural selection can however give us abilities that may be unrelated to those originally selected. The brain may have developed the ability to make and prune many random synaptic connections in childhood because this increased our survival probability in a new and changing environment. This has also enabled us to learn things that may not be perceived to be biologically useful.

We will see in chapter 4 that the human brain is endowed with another amazing capability in that it can generate its own information or memory-like states, which were not intentionally stored in the network. These so-called spurious states are made up of combinations of features of stored memories, in endless variety. They may correspond to what we commonly identify as creativity, or the ability to do something new or different. Spurious states clearly give the brain added flexibility to adapt and to develop nonbiological abilities.

FURTHER READING

Principles of Neural Science, edited by Eric R. Kandel, James H. Schwartz, and Thomas M. Jessell. Englewood Cliffs, N.J.: Prentice-Hall, 1991. This textbook is recommended for serious neuroscience students. It contains a host of information on how the brain is organized and how the various components of the brain work together.

The 3-Pound Universe, by Judith Hooper and Dick Teresi. New York: Jeremy P. Tarcher, Perigee Books, 1992. An excellent, very readable, and thorough book about the human brain in general, written by two science writers who spent three years traveling around America interviewing experts in the neurosciences. This book covers a broad range of topics and gives a good account of some of the history behind the neurosciences.

Drugs and the Brain, by Solomon Snyder. Scientific American Library. New York: W. H. Freeman and Company, 1986. An excellent Scientific American Library book, by an expert in the field, which reviews what is known about the chemistry in the brain.

Neuronal Man: The Biology of Mind, by Jean-Pierre Changeux. Oxford: Oxford University Press, 1986. Reprinted Princeton: Princeton University Press, 1997. Gives a very thorough and interesting account of how the nervous system of animals and humans may have evolved and how it may develop from the fertilized egg to the mature human brain.

CHAPTER 3

The Remembering Brain

Memory is the process (and the information stored by that process) that enables us to carry our experiences into the future, and to look back into our past and use those experiences. It clearly offers enormous biological advantages, and that is why all animals (and some plants) have some form of memory. In animals the main form of memory is a function of the nervous system.

Memory allows us to react quickly to known stimuli and to avoid making the same mistakes as before (and to "learn" from our mistakes). With memory we can refine our methods and techniques by taking shortcuts to avoid superfluous and useless steps. Memory allows us to avoid dangerous situations and predators, but at the same time helps us to gain an advantage over our prey. It helps us remember where to find food, how to gather food, and what foods are safe (not poisonous or too dangerous to catch). In humans, memory is enhanced by language and extends beyond the biological brain. Memory or information can be stored in books, computers, and on the Internet, and it can be shared with other people. This gives us additional informational capabilities that have enabled us, among other things, to understand the world around us, because we can build upon our knowledge and work collectively as a society.

Nervous-system memory is not the only form of memory. Our immune system, for example, keeps a memory of which bacteria have recently attacked our bodies, and how to deal with these intruders more effectively in the future. Plants also have a very basic form of memory, which is probably driven by simple chemical reactions. They "know" when to flower, when to lose their leaves, when to fruit, when to send out new shoots, and how not to be tricked by unseasonable weather conditions. They know to open their flowers during the day and close them at night (to conserve their perfume for the insects) and to open their leaves during the day (to absorb the sunlight) and close them again at night (to reduce water loss). Much of this memory has genetic origins, or is programmed memory, but it is nonetheless a form of memory. The nervous system gives an animal the genuine capacity to learn something new from its environment, something that its parents may have not encountered.

Nervous-system memory can be recalled quickly, giving animals the ability to respond rapidly to (known) environmental stimulation. Animals can also store many more memories than plants. The most important characteristic of nervous-system memory is that it is malleable. This allows animals to adapt to new situations. Animals are able to alter a response so that it is more economical and efficient. In humans, memory has an almost unlimited capacity, and new memories and ideas can be created and altered in the mind (at will). This is what we do when we think and speak, or when we are creative. Memory gives animals an enormous advantage over plants. You may wonder why plants exist at all, given that animals have such a great advantage. The key survival characteristics of plants are that they are generally much more prolific and some of them live for much longer than animals do. (Insects are generally quite prolific for the same reason: they have simple nervous systems.)

Neural memory is very different from memory stored in a computer or in a recording device. It is not a faithful representation of the facts and it can change with time, and the recall of memory is a constructive process, as opposed to computer memory, which is recalled literally. In this chapter I present a qualitative description of how memory works in the brain, based on neural network models, that captures some of the salient features of neurobiology.

CLASSIFICATIONS OF MEMORY

Our main concern in this book is with human memory, although humans do share many features of memory with other animals. In fact, much of what we know about memory and how the nervous system works has resulted from studying simple animals like the sea slug *Aplysia* (which has only twenty thousand neurons) and the fruit fly, and mammals like mice, rats, cats, monkeys, and chimpanzees. (Chimpanzees are especially interesting because they are closely related to humans. They even seem to have self-awareness and demonstrate social behaviors similar to those of humans, such as male gang/boardroom mateship and casual sex.) Animal studies have been useful, since the best we can do with human experiments is to study patients with brain damage (which may have been caused by an accident or resulted from brain surgery) or use noninvasive imaging techniques to look at the brain while it is functioning.

There are a number of very famous patients, one of whom was already mentioned in chapter 2, who have helped advance our knowledge of the brain. H. M. had his hippocampus, and some parts of his temporal cortex located around the hippocampus, bilaterally removed to control epilepsy (Scoville and Milner 1957). Although H. M. appeared to be normal—he could read, write, reason, and carry on a conversation—his memory span did not extend beyond fifteen minutes or so. He quickly forgot what he was talking about if he was distracted. H. M. was unable to lay down medium- or long-term memory. If he had met you earlier and you walked out of the room for fifteen minutes, he would not remember who you were if you were reintroduced to him. This demonstrates that the hippocampus and surrounding areas of the brain are essential for storing memories away more permanently. H. M.'s form of amnesia also reveals that there are different types of memory, namely short-term (or working memory) and long-term memory.

H. M.'s case suggests that the hippocampus (and perhaps parts of the temporal cortex) is required to access fairly recent and older memories in the neocortex, or possibly to collect together the various components that make up a memory. The other strange thing about H. M. is that he still had all of his much older memories, which he acquired some three years prior to his operation. (That means that immediately following his operation, H. M. lost all of his memories that were less than three years old.) This suggests that the hippocampus is involved in slowly converting memories into more permanent storage, where eventually they no longer rely on it. This process may take around three years to complete, and without the hippocampus and proximate structures in the temporal cortex, it cannot happen.

It was discovered later by Brenda Milner (the psychologist who first studied H. M.) that H. M. was still able to lay down some long-term memories. He could learn, for example, how to trace an image in reverse by watching his hand drawing the image in a mirror. His ability to do this improved with each new attempt. H. M. could also solve the "Tower of Hanoi" puzzle and improve on each previous effort, although he thought he was solving it for the first time every time he attempted it. He could still lay down some motor and skill memories. This means that such memories are processed through channels that do not involve the hippocampus.

These discoveries suggest another classification of memory, in terms of so-called declarative memories (which are generally conscious, reflective, thinking, and explicit) and nondeclarative memories (which are automatic, reflexive, unconscious, nonthinking, and implicit) (Squire and Kandel 1999). Declarative memories are memories we bring to mind and become conscious of, whereas nondeclarative memories are memories like a skill, or a motor action, that does not need to enter consciousness.

Conscious or declarative memories can be further classified into semantic and episodic memories. Semantic memory refers to facts, like who was the first president of the United States. Episodic memory refers to personal memory, which involves us directly, such as where we went on holiday last year, what we ate for breakfast this morning, or what we said to someone earlier today. The type of declarative memory most affected by H. M.'s amnesia is episodic memory, which suggests that the hippocampus is primarily involved with the long-term storage of episodic memory. Semantic memory may not have to go directly through the hippocampus.

Just as the hippocampus is required to lay down long-term episodic memories, the amygdala (located close to the hippocampus) is involved with the storage of emotional memories, or the emotional aspects of memories. People with damage or lesions to the amygdala can recall only those parts of a particular memory that do not have any emotional content. Other small brain organs are responsible for other types of memory, such as skill and motor memories.

It is known that emotions and moods play a prominent role in learning, or the process of laying down memories. Consciousness also plays an important role in our learning experience. We seem to have to become conscious of something before we can learn it properly. This does not apply to all forms of memory, as some things can be learned without entering consciousness. But it is thought that even most nondeclarative memories may have at one time been conscious memories. When we first learn to ride a bicycle, we have to think about what we are doing, but then with sufficient practice, familiarity, and dexterity, this memory is converted into an unconscious nondeclarative memory. We are then able to carry out the appropriate actions without much recourse to the conscious mind. The more often something is presented to us or encountered, the greater is the trace of that memory. In fact it seems that something may need to enter consciousness a few times, or that we need to rehearse it a few times, before it enters long-term memory. This process is what we call on when we study for examinations.

As pointed out previously, memories can be classified according to how long they last: a few seconds, a minute, a day, a week, a year, or a lifetime. The shortest form of memory seemingly concerns the moment that just passed or was experienced. This form of memory is thought to be intimately linked to our conscious experience (James 1890). Another form of short-term memory, called working memory, lasts a little longer, from a few seconds up to a few minutes. Working memory is essential for us to have a conversation, to read something, to comprehend what we are reading, or to remember a telephone number, and is synonymous with the famous limitation to 7 ± 2 items (Baddeley 1990; Miller 1956). These forms of short-term memory can be classified further, but beyond fifteen minutes or so we generally head into the territory of medium- and long-term memory.

Long-term memories can also be classified. Some last for days, others last for a week, and some last much longer. Some permanent memories are probably hard-wired from birth (such as how to breathe and how to control our cardiovascular system) or in the early infancy and childhood development phase (such as how to walk, how to eat, and how to speak).

Besides the very short term memory, which is thought to be intimately linked to our conscious experience, there is an even shorter form of memory that we are not directly conscious of. If we are shown something for a very short period, too short for us to become conscious of what we were shown, the brain (or the unconscious mind) still seems to remember something about what it was shown (Sperling 1960). Experiments reveal that subjects have unconscious knowledge of what they were shown. In the 1960s, it was common practice to flash subliminal messages on the movie screen just before an intermission, to get people to buy certain products during the break. These images would typically last for a twentieth of a second or less, short enough so they did not enter consciousness, but long enough to be processed by the unconscious mind. Drink and candy companies exploited this unconscious form of short-term memory.

In addition to memories that are "intentionally" stored in the brain, there are memories generated in the brain itself, because of the way that memories are stored in a distributed and overlapping fashion, sharing neurons and synapses. These "spurious memories," which are observed in mathematical models, are generally considered to be a nuisance. They may explain why we cannot remember certain things at certain times, why our memories become distorted with time, and how "false memories" arise. Spurious memories, however, may also be responsible for our creative ideas, and may be what actually allow us to think, adapt, generalize, and learn. These memories and creativity in the nervous system are considered in chapter 4.

In the following sections I address the question of how neural networks facilitate the storage and retrieval of memory.

WHERE IS MEMORY STORED IN THE BRAIN?

Memory is thought to be stored in the brain as certain stable patterns of highly active, persistently firing neurons, much like a picture made of black and white pixels, with the white pixels corresponding to active neurons and the black pixels to quiescent neurons. A memory is recalled (or formed; see below) when the particular pattern associated with that memory reverberates for a short but extended period of time, on the order of ten milliseconds or more.

Historically, it was originally thought that memories were stored in specific locations in the brain, involving small numbers of neurons. According to this theory, memories were housed in (almost) mutually orthogonal compartments, in much the same way that memory is now stored in a computer. An extreme version of this theory suggested that a memory is stored in each neuron. This came to be known as the "grandmother cell" theory, because in this theory the memory of one's grandmother would be stored in a single neuron. Detailed experiments in search of the memory engram (or specific location of a memory) failed to validate these extreme hypotheses. Karl Lashley (1890–1958) died after a fruitless lifelong search for the memory engram in mice that he had trained to find food in a maze. Little by little Lashley systematically removed bits and pieces of the cerebral cortex of these mice but was unable to identify where the memory to navigate the maze was stored in their brains. What followed was a period when scientists thought that memory was stored in the whole brain, or almost all of the brain, widely distributed over large areas. This would explain the negative results of Lashley.

The actual truth, though, is somewhere between these two extremes, leaning perhaps more toward the specificity of location of memory (Churchland 1986; Hooper and Teresi 1992; Squire and Kandel 1999). Incidentally, researchers have recently been able to find the trace of a particular memory in the brains of mice (Squire and Kandel 1999). The problem was that the particular memory Lashley was looking for had a number of different components to it and consequently was simultaneously stored in a few different places in the brain. Different groups of these components could reproduce most of the maze memory. Lashley also confined his attention to the cerebral cortex, and it is now known that other structures in the brain, such as the hippocampus and the amygdala, may also be important for the storage of memory, particularly the hippocampus when it comes to navigation memory.

In summary, memories are thought to be stored in the brain in a distributed manner, but in semispecific areas. Visual memories are primarily stored in the visual cortex (as well as some other interior substructures), auditory memories are stored in the auditory cortex, and associations are stored in the associative cortices. More detailed visual memories are stored in specific locations within the visual areas, and similar memories are probably stored in the same general area or alongside each other, as this enables the brain to "compare" them easily.

Lashley's experiments, as well as the examination of patients with brain damage or lesions, demonstrate the fact that memories are robust. The robustness of memories is a natural consequence of the distributed storage of memory. If a memory was stored in a single neuron only and something happened to that neuron, the memory would be lost forever. But since memories are stored in groups of neurons (say on the order of a thousand or ten thousand neurons), it does not matter if a few neurons die or if they are not initially activated by an input. We will see later, when we look at a specific model, that neurons in the correct firing mode for a particular memory will activate the other neurons that are initially not activated by the input. In other words, neurons act collectively to recall a memory. This is what gives memory its robustness and allows us to recall memories with imperfect or partial input. When the brain is presented with an input that closely resembles a stored memory but is not identical to it (like the faces of our family and friends, which change from day to day), the majority of the neurons that are in the correct firing mode will help to excite those neurons in the memory that are inappropriately quiescent. (Neurons that are improperly firing will also be quelled by the general inhibition and lack of support they will subsequently receive from the neurons that continue firing.) When we want to recall something, a piece of knowledge or an episodic memory, we normally put some cues (or clues) concerned with the sought-after memory into our mind and let the brain search its memory store for the most appropriate memory corresponding to that input. This ability is a function of the robustness of memory.

We saw in chapter 2 that the neocortex and most of the brain are thought to be organized into modules, or smaller neural networks, each containing somewhere around a thousand to ten thousand neurons (Amit 1995; Changeux 1986; Churchland and Sejnowski 1992). The neocortex, for example, has process-specific columns and microcolumns, and most other brain structures are also organized in subgroups called nuclei. In the neocortex, some of these modules are involved in specific processing of sensory information, such as extracting vertical or horizontal edges from the visual field. The modules are not clearly distinguishable from each other, as they are also connected to each other. In the case of cortical columns, there are many lateral connections between columns. Modules are generally identified by a higher local intensity of connections among groups of neurons. The pattern of network connectivity (how it is organized in modules) is also a reflection of the limitations in the physical space available inside the brain. The pictures in figures 2.3 and 2.4 show how dense and intricate the connectivity actually is.

Although memories are thought to be stored in semispecific locations, a complete memory may contain aspects that are stored in a number of modules in different areas of the brain. For example, a memory of a particular red racing car may involve a number of visual aspects (the color red, motor car, racing) stored in different areas of the visual cortex; the sound of an engine running (idle and racing) stored in the auditory cortex; an emotional aspect (like love or fear, or both) stored in the amygdala and other parts of the cerebral cortex; the name or the make of the car, stored in the language (and speech) areas of the brain; and a connection to other memories (such as a particular racetrack, racing event, or racing driver) stored via the association cortex. Most of these aspects of the memory are recalled "simultaneously" when the memory comes to mind.

In this paradigm, a complete specific memory is probably stored in a few modules, on the order of ten or so, with each module storing a particular feature of that memory, such as its color, shape, size, an associated emotion, and a time. A memory is recalled when all of these modules, which are relevant to that particular memory, are simultaneously activated. Just as neurons can correct each other, so can modules. If we input only some of the features of a memory (each of which is stored in a different module), the modules in the correct firing pattern (corresponding to that stored memory) will excite the modules that are initially not in the correct configuration for the memory. If a sufficient number of modules are in the correct state for a particular memory, they will work together to excite the other modules that carry the other features of that memory, and the complete memory will be recalled. In this way, features of memories help to collectively excite each other.

Particular module configurations can be shared among different memories, just as different memories, or memory features, can share the same neurons. The "red" state in the color module is shared by the memory of a "red car" and a "red rose," for example. If we were to input the cues "red" and "perfume," we might recall the memory of a red rose. Memories that share common module states may either be in competition with each other (if their overlap is small) or they may excite each other (if their overlap is large).

In this representation, the image of a rose in one module may excite the smell of a rose in another module. In some cases these modules can be far apart. This mutual excitation is probably facilitated by long-range excitatory interactions. It is interesting to note that most neurons with longer projections (such as pyramidal neurons) are excitatory neurons. Similar memory states, or memory features (like different makes of cars), are probably stored close to one another in the same or neighboring modules. This would facilitate the task of distinguishing between them. Competition among neighboring modules, and within the same module, is facilitated by inhibition. It is interesting to note that the majority of local-circuit neurons are inhibitory (Crick and Asanuma 1989).

When a collection of modules, corresponding to a memory, are simultaneously activated (in stable, active configurations) for a prolonged period of time (of around a few tens of milliseconds), they are somehow bound together to form the perception of a complete memory. Exactly how this happens is not known, but it seems to involve the attentional aspects of consciousness. Some possibilities are that special awareness neurons (maybe involving a special brain organ like the hippocampus or thalamus) observe what is going on in the other parts of the brain; or perhaps the set of firing neurons somehow synchronize their own activity and observe each other in some way.

The storage of memories in more than one neuron in a module, and in more than one module in the brain, has a certain robustness on two levels. Each system (at the neural or the module level) can collectively correct for noisy inputs at the appropriate level. If one neuron is in the wrong firing mode, the other neurons in the module will correct (for) it, and if one module property of a memory is incorrect, then the other modules will correct for that too. In the theoretical model, to be discussed later in this chapter, some memories can still be recalled if as many as 30 to 40 percent of the neurons (or modules) are in the incorrect firing mode. This is quite remarkable because many memories are simultaneously stored in these neural networks.

Neural memory is sometimes also referred to as "associative memory." The term was originally coined by the famous Russian psychologist Ivan Pavlov (who made dogs salivate when they heard a bell), referring to the association of two or more different memories together. One can also apply this terminology to the association of features of a single memory. Associativity works as a result of the way different memories or parts of the same memory are linked together in the brain. This can happen if memories have common features or if they occur at the same time. I will suggest in chapter 4 that spurious memories may themselves also provide a basis for making associations between memories. This is because spurious memories combine aspects of memories together and, in a sense, represent a commonality among the different memories they are constructed from.

Incidentally, association is a good practical method for remembering something. If we associate one memory with another, this seems to help us later when we want to recall the original memory. By making associations among memories we are laying down more paths for the associated memories to be recalled later. Associations provide more cues to recall a stored memory. Context is also an important form of association. Sometimes we see people whom we (should) know out of context (they may not be wearing their work uniform, or we may see them at the beach), and we cannot recognize them immediately. Context is in a sense part of the complete memory, even though we may not be directly conscious of it being so.

THE BRAIN VERSUS THE COMPUTER

In a computer, memory is stored as a binary number, and associated with this number (and every other memory stored in the computer) is another binary number that tells the computer, the program, and the programmer where exactly this memory is stored. This is very different from how memory is stored in the brain, where it does not have a specific address but is recalled from imperfect input. It is said to be "content addressable" because it is recalled from the content of the input, and not from an address, as in the case of computers. This is how we are able to recognize people we "know" even though they almost certainly do not look exactly the same as when we last saw them. A conventional computer would have great difficulty with such a task. Computers are inflexible. They need a precise address to recall a memory and cannot handle any errors or noise.

Human memory also changes with time. Some of this change is caused by the decay of the neurotransmitter that is stored in the presynaptic vesicles or by the fact that memories interact and interfere with each other, due to their storage in common areas. Computer memory stays intact, as it is stored in separate locations using less perishable elements. The rigid storage of computer memory, however, means that if something goes wrong, then all is lost. In contrast, memory storage in the brain is quite robust.

If a single computer instruction or address is incorrect, the computer will cease to execute the program or will be unable to recall the correct memory, whereas in the brain and in neural networks generally, noise (or incorrect bits of information) does not prevent the recall of the most suitable memory corresponding to an input. Neural networks have a built-in error-correcting mechanism based on the fact that cells and modules work together. This is a good thing, as noise is a widespread phenomenon. We will see later that neural networks may actually even work better in a slightly noisy environment. The internally generated spurious states, mentioned earlier, generally interfere with the retrieval of stored memory. It turns out that noise helps a neural network avoid getting stuck in some of these spurious memories in the process of recalling a stored memory.

Today's computers are quite fast. A personal computer (PC) can carry out over 100 million instructions per second (or 100 MIPS, in computer jargon). Neurons, on the other hand, are relatively slow. They execute an instruction (actually all they do is fire, or emit a pulse) every few milliseconds at best. (After a neuron fires, it cannot fire again until a refractory or rest period of a few milliseconds has expired.) This means that a neuron can execute only a few hundred instructions per second, which is much slower than a PC. One of the main reasons this is so is that neurons rely on the transfer of chemical currents across the synaptic cleft. Neurons also need to receive sufficient simultaneous excitatory input from a hundred to a thousand other neurons before they fire.

Given all of this, the brain is still much faster than a computer when it comes to certain tasks, like pattern or voice recognition. The key difference is that the brain works in parallel, as all of the neurons work simultaneously, whereas in a computer, the processing is serial, and only one instruction can be executed at a time. (A computer may have several processing units, but generally speaking it is serial.) This situation should be compared to what is going on in the brain, where theoretically hundreds of billions of neurons are working at the same time. If one takes this into account, the computational capabilities of the human brain may be on the order of ten million MIPS (or ten million million instructions per second), which is a hundred thousand times more powerful than a PC. What the brain and neurons lose in speed is made up by having a large number of cells working together at the same time. When the brain performs a computation, like the recognition of someone's face, each neuron (that does fire) probably fires a few times before a specific stable pattern of neural firing activity emerges and the memory is recalled. Based on this, the recognition of someone's face may take a few hundredths to a few tenths of a second. The computer, however, is faster at performing serial, repetitive tasks, like multiplying numbers or manipulating data. These tasks are vastly different from the sorts of things the brain is good at.

In a later section of this chapter, I elaborate on the storage capacity of the human brain.

ATTRACTORS

One of the most influential ideas concerning brain function is the notion that a memory is recalled when a certain group of neurons reverberate together in the same state of excitation. In this state (called an attractor), only the same set of neurons keep firing repeatedly. The idea of attractors (or reverberations) dates back to the remarkable genius of the psychologist Donald Hebb, who came up with a proposal of how collections of neurons could learn by adjusting the synaptic connections between them (Hebb 1949). Attractors are the foundation of some of the most interesting memory models today. They were first noticed in computer simulations as early as 1956, shortly after Hebb proposed his hypothesis (Rochester et al. 1956). The idea did not gain widespread recognition until after the biophysicist John Hopfield gave the memory recall process a physical interpretation (Hopfield 1982a).

In general, a dynamical system (a complex interacting many-body system that evolves with time) can, depending on the input, converge to a fixed-point state (or an attractor), a limit cycle (or periodic orbit), or a "strange attractor" (or behave chaotically). Attractors are states of the system that are stable and, once attained, are usually maintained for a long period of time. In limit cycles, the state of the system changes in a cyclic manner, and the same sequence of states is repeated after some time. An example of a limit cycle is the weather repeatedly going through the four seasons—winter, spring, summer, autumn. The problem with limit cycles, in dynamical systems theory, is that they are generally unstable, since they require fine-tuning of the parameters in most models to maintain stability. A small perturbation usually sends the system off into another quite different state. It is unlikely that the brain would make much use of these states for computation.

The third possible behaviour of a system, the approach to a strange attractor, is similar to a limit cycle, except that the system does not exactly come back to one of its previous states, but close to it. The weather is actually a strange attractor as the system does not really return to exactly the same configuration. Taking temperature, humidity, rainfall, and the like into account, one finds that the weather is never exactly the same year after year. Strange attractors are a hallmark of "chaos theory," or systems that are eventually sensitive to errors and noise. Although strange attractors are not formally stable configurations, they do possess more stability than limit cycles. This has led some researchers to suggest that they may be useful to the brain (Freeman 1991; Yao and Freeman 1990). The discussion in this book, however, is based on the premise that attractors are the mainstay of cognitive significance. Attractors are a much more practical alternative to represent a significant cognitive event, such as the recall of a memory or the reaching of a decision.

In this framework, a memory is recalled when an ensemble of active neurons collectively sustain their mutual activity at an elevated level. These states of stabilized activity are fixed-point states (or attractors) in the neural dynamics. When the brain converges to one of these memory states following an input, it will remain there for a short period of time, until it is destabilized by the next input. This drives the system into another attractor, or memory state.

Daniel Amit, one of the pioneers of "attractor neural networks," goes so far as to suggest that attractors in modules may constitute the basic building blocks, or the "atoms," of brain function (Amit 1995). They are an extremely logical choice to represent the retrieval of memory, or the act of reaching a decision. When an attractor is attained, it signifies that the input was familiar, so this also represents the act of recognition. An attractor also serves as a temporary store of a stimulus after it has been removed (short-term memory) and provides a separation point for the complex connection between sensory input and motor output. As we will see in the next section, an attractor provides a basis on which learning can take place. Attractors are also interesting because they are impervious to noise, and to the way that real neural networks are randomly connected.

As Amit points out, however, no matter how appealing attractors may be, we still need to observe these formations experimentally in the brain if they are to be taken seriously. Some evidence for attractors, or local reverberations, comes from delay activity experiments conducted by Yasushi Miyashita and his coworkers (Amit 1995; Fuster 1973; Miyashita 1988; Miyashita and Chang 1988; Sakai and Miyashita 1991). In these experiments, monkeys were presented with uncorrelated pairs of images and were later tested to see if they could recognize the matching pairs of patterns they were shown. To encourage the monkeys to do their best, they were rewarded if they got the correct answer. During the testing stage, each monkey's brain was monitored with electrodes in the anterior ventral temporal cortex, where it was thought that portions of these images or memories might be stored. The result was that reverberations (or attractors) were observed for matching pairs of images in the monkey's brain, which persisted for up to sixteen seconds after the stimulus had been removed. These reverberations were different for different stimuli, but reproducible for the same stimulus.

LEARNING

The process by which memory is actually stored in the brain, as we have seen, is the production, storage, and release of neurotransmitter in and from vesicles in the presynaptic neurons. At the synaptic clefts, neurotransmitters, with their partner neuroreceptors, control the amount of neural current that flows across. Previous firing patterns (or simultaneous correlated activity among neurons) are recorded as increases in the synaptic efficacies in such a way that in future, neural currents flow with greater ease in those channels. This helps to reestablish the transition to a previous memory, which caused this change in the first place. This is the essence of Hebb's hypothesis: "When an axon of cell A is near enough to excite a cell B and repeatedly or persistently takes part in firing it, some growth process or metabolic change takes place in one or both cells such that A's efficiency, as one of the cells firing B, is increased" (Hebb 1949). This is the generally accepted view about how learning takes place in neural networks. Repeated use of a particular synapse results in more neurotransmitter being made available for release from the presynaptic vesicles. If there were no synaptic clefts and the neural currents flowed uniformly, unimpeded and unaltered, it would not be possible to store memory in the brain in a nontrivial way.

If a neuron is persistently active, the synaptic efficacy is strengthened at the input connections that are exciting it and slightly weakened at those connections that are not exciting it. If the postsynaptic neuron is not firing, there is generally no change at the incoming synaptic clefts. How the postsynaptic cell communicates back to the presynaptic cells is a bit of a mystery, as the flow of information is generally in the opposite direction. One suggestion is that when a postsynaptic cell is repeatedly excited it releases chemicals like nitric oxide (NO), which drift back to the presynaptic cells and cause them to prepare more neurotransmitter for future use. The presynaptic cells presumably respond only if they are themselves active when the NO reaches them. In real neural networks, learning generally takes place only on excitatory synapses or neurons.

The learning process alluded to above probably takes place in a largely unsupervised manner. Whenever a neuron is repeatedly excited, the connection is strengthened to those neurons that were exciting it. There are some arguments to suggest that supervision or observation (by other brain organs) may be taking place too. Fragments of attractors need to be combined in some way (called the "binding problem") to form a complete memory. The hippocampus may be involved in this function. Another particularly interesting candidate is the thalamus, which is the gateway to all sensory information entering the neocortex and for some reason also receives reciprocal information from each of the areas in the neocortex that it sends information to. It is also known that neurotransmitters like norepinephrine are implicated with learning and are thought to be released from the locus coeruleus when something is to be learned. This needs to be coordinated with the task of observation. As mentioned above, learning seems to be enhanced if our emotions (such as fear or joy) are heightened. Presumably the amygdala sends signals to the locus coeruleus and other related brain structures to release more neurotransmitters, which help the brain to cement the learning experience.

Memory recall and learning take place simultaneously. When an attractor is attained, this represents the recall of a memory and the opportunity for learning. In an attractor state, learning takes place by changes in the appropriate synaptic efficacies, as outlined above. New memories stored in the network become part of the machinery that processes future input. Memory and learning are complementary processes. What a network learns depends on its synaptic structure (or previously stored memory), as this determines which attractors are attained and learned. Put another way, we perceive the world according to our own memory store, and this influences our future perceptions.

The general scenario suggested above is that the brain processes information and, if the input has some relevance the neural system will approach an attractor. When an attractor is achieved and the same set of neurons fire persistently, Hebbian learning takes place. This process then installs new attractors into memory. There is a bit of a dilemma here in that new attractors must already be in the network. As noted previously, when we store memories in neural networks in a distributed overlapping fashion, one of the consequences of this is that the network generates its own set of memories that were not intentionally stored in it. These spurious states are very weak memories or attractors, and I suggest in chapter 4 that they are required when something new is to be learned.

Neural networks have double dynamics. The neurons are updated on a time scale of a few milliseconds, corresponding to the refractory period of a neuron. This dynamics is driven by the synaptic efficacies, and most of the attractors arrived at in this process are related somehow to the memories (or combinations of these memories) that are already stored in the brain. Learning or synaptic dynamics takes place on a time scale of tenths of seconds, which corresponds to the time it takes a network to reach an attractor and reverberate in that state.

In this model, learning takes place whenever a neural network reaches an attractor, so that the more times a particular attractor is reached, the more times and stronger that attractor will be encoded in synaptic memory. This agrees with the well-known fact that repetition and rehearsal enhance learning.

A limited amount of neurotransmitter can be produced at the presynaptic cells, so one expects that the synaptic efficacies cannot be increased indefinitely. In other words, there is a limit to how large the synaptic efficacies can become. It is also thought that neurotransmitter is released in certain quanta (bundles of five thousand molecules) from the synaptic vesicles, so the allowed values of the synaptic efficacies are also quantized (or take on discrete values that differ by a set amount).

An interesting point that is often mentioned by people who use drugs is that they seem to have a different set of memories depending on their state of mind, that is, whether they are under the influence of a drug or not. (There is some experimental evidence to support this.) This suggests that people may lay down different memories when a different neurochemistry (induced by a drug) is operative in their brain, and these memories may not be retrievable unless that same neurochemistry is attained again. In terms of attractors, this means a different attractor structure may exist for different neurochemical states of the mind. This also explains why our perception is altered when we are under the influence of drugs. We may see the same situation in a completely different way, depending on our state of mind at that time. Emotions and moods also affect our perception and have an underlying neurochemical basis. We know this because certain drugs can change our moods and emotions.

Another example of this is the difference between waking and dreaming consciousness. Each of these states of mind is controlled or influenced by a different brain chemistry. We seem to have a different set of memories when we are dreaming compared to when we are awake, but this perception may be a result of the delusional nature of dream sleep. We often have dreams in which we feel we know some of the fictional characters from a previous dream. Sometimes when we wake up briefly and go back to sleep we slide back into the same dream, with the same characters. What is interesting about this is that during dream sleep the brain is influenced by a different set of neurochemicals than when we are awake. During dream sleep the brain is cholinergically driven, while in waking consciousness it is driven by aminergic neurotransmitters, like norepinephrine and serotonin. It would seem from these examples that the chemistry of the brain plays a major role in what we recall and how we perceive events around us.

This scenario may also have something to say about "multiple personality disorder," a syndrome in which someone appears to have multiple personalities or identities. In this case the separation of memories is so strong that it seems to affect the "self" of each of these individuals. Multiple personality disorder may have a neurochemical basis, with different disjointed memory stores. People with the disorder usually have no recollection of what they may have done when they were someone else.

TEMPORAL MEMORY

Attractors may also provide a basis for the storage (and retrieval) of so-called temporal memory, such as the recall of the alphabet, a tune, a poem, a movie sequence, an event in sport, or even the coordination of a physical movement.

Such memories can be stored as a sequence of static memories or attractors, with one attractor leading to the next and so on (Kleinfeld 1986; Sompolinsky and Kanter 1986). The process by which these types of memories can be stored in a neural network is quite natural. When a new stimulus arrives at a network while the network is still in an attractor state corresponding to the previous input, this destabilizes the network, which then starts to process the new input. This leads to another attractor, which is in turn destabilized by another input, and so on. During these transition periods, when an attractor is being destabilized, the network is simultaneously holding a representation of that attractor or memory while it is also starting to process the new input. Older attractors may also still be reverberating weakly, while other attractors are being achieved. In this way, the network can easily learn correlations between (almost) simultaneously existing states, by a simple Hebbian-type process.

Later on, when the network finds itself in one of these attractor states, it will automatically become destabilized as a result of this learned correlation between concurring brain states, and the system will be driven to the next attractor, followed by the next, and so on. In this way the network can learn to recall a sequence of static memories or attractors. This is one of the reasons why it is easy to recall the alphabet in the usual order, but much more difficult to do it in the reverse order. The forward process is naturally programmed into the network by a process as outlined above, but reciting the alphabet in reverse requires much more effort. In fact, the task of reciting the alphabet in reverse order is usually accomplished by running through the alphabet in the usual order, for groups of three or four letters at a time, and noting which letter comes before the letter we are currently at in our reverse citation.

The delay activity experiments also provide evidence that the brain is able to store memory in this way. Miyashita observed correlations between the recorded neural activity patterns for consecutive images, even though the inputs were uncorrelated in appearance. In effect he found that the monkey's brain had converted temporal correlations (things presented to it in sequential order) into spatial correlations in the brain. Miyashita observed distinct correlations in the neural activity patterns for the last five or so images presented to the monkey (Miyashita 1988). This is interesting because theorists claim that the same number closely fits their expectations from attractor neural network models (Griniastry, Tsodyks, and Amit 1993; Amit, Brunel, and Tsodyks 1994).

Attractors are destabilized every few hundredths of a second. Although there is the capacity for the brain to move between attractors in different parts of the brain, this form of temporal memory does not seem to be sufficiently brisk to explain the capabilities of the human brain to deal with such things as playing music. This type of processing may be enhanced if the nervous system somehow uses the timing of action potentials arriving at neurons as a computational tool, or takes advantage of the synchronization of neural firing rates (Hopfield 1995; Hopfield and Brody 2000, 2001).

Temporal sequences made up from sequences of motionless attractors involve static memories that are very similar to each other. Highly correlated memories are naturally and strongly associated with spurious memories (which naturally encode the correlations between memories), and one way the dynamics of temporal memory may be accelerated is if these spurious memories somehow act as filler between consecutive static memories. In this way the brain may need to do less processing to represent the same temporal sequence. This idea may explain why our temporal memories appear to be continuous.

THE HOPFIELD MODEL

Interest in neural network models as a means to understand memory and brain function soared after John Hopfield came up with a very simple theoretical model that captured the basics of content-addressable memory. The Hopfield model gave the process of memory recall a physical interpretation in terms of an "energy" minimization process and was amenable to mathematical analysis (Hopfield 1982a, 1982b, 1984a, 1984b). In this model, memory is represented as attractors or fixed-point states in the neural dynamics. Although the Hopfield model makes several simple and unrealistic assumptions about neurons and their connectivity, some of which are considered drastic by the neuroscience community, the general results of the model are accommodating to the inclusion of more realistic biological features (even for quite radical alterations), and furthermore such changes lead to even more realistic results. For these reasons the French physicist Gerard Toulouse coined the phrase "a clever step backwards" to describe Hopfield's idea (mentioned in Gutfreund 1990). More sophisticated models, which incorporate more realistic biological features into the Hopfield model, are collectively referred to as "attractor neural networks," since in all of these models attractors are the basis of cognitive significance (Amit 1989; Peretto 1992).

Figure 3.1. Dr. John Hopfield, a biophysicist from Caltech (now at Princeton University), put forward the idea that memory may be stored in the brain in terms of attractors, or fixed-point states in the neural dynamics, in which the neural network reverberates in a constant state of excitation. His seminal paper in 1982 (listed in Further Reading) led to a revolution in neural network research that continues today. *(Photograph courtesy of Denise J. Applewhite.)*

In the "naive" or simplest version of the Hopfield model, all of the neurons are completely and symmetrically connected to each other. Each neuron can take on one of two possible values, $+1$ (corresponding to a rapidly firing neuron) or -1 (corresponding to a quiescent neuron, or one firing at background levels). In this model, a memory is represented by a certain pattern of $+1$s and -1s for the activation states of the neurons in the network. An example of such a memory is shown in figure 3.2 for a network with a hundred neurons. Here the $+1$ state is represented by a white square cell, and the -1 state is represented by a black cell.

A memory is stored in the network by a generalized symmetric form of Hebb's rule. If two neurons are in the same state of excitation, either both firing $(+1)$ or both quiescent (-1), the synaptic efficacy between them is strengthened by one unit, and if two neurons are in opposite states of excitation, the synaptic efficacy between them is weakened by one unit. Neurobiologically, a synaptic efficacy is normally increased only if a postsynaptic cell is firing concurrently with a presynaptic cell, and possibly decreased if the presynaptic cell is quiescent while the postsynaptic cell is firing. No change takes place when the postsynaptic cell is not firing, and certainly not when both the presynaptic cell and the postsynaptic cell are not firing. Yet in the Hopfield model the synaptic weights connecting two quiescent neurons in a memory are upgraded. In real neural networks, neurons are also not symmetrically connected, and memories are not stored in such a symmetric fashion.

The unrealistic learning rule used in Hopfield's model is specifically chosen to maintain the inherent symmetry in the model, which is important to make the initial connection of memory with minimum "energy" states and fixed-point attractors. Figure 3.2 illustrates how some of the synaptic weights in this model are strengthened or weakened to enable the network to learn the pattern shown. Each link represents the synaptic efficacy for either neuron affecting the other it is connected to. In this case, the memories (or patterns) are stored in the network "by hand"; in more realistic attractor neural network models with double dynamics, however, the network can be made to learn its own attractors. This unsupervised process is probably more in line with what happens in a real neural network.

In general one can store many different patterns in the same network. In the Hopfield model the storage capacity is on the order of 0.14 times N patterns, where N is the number of neurons in the network. This means that in a network with a hundred neurons, one can store about fourteen patterns similar

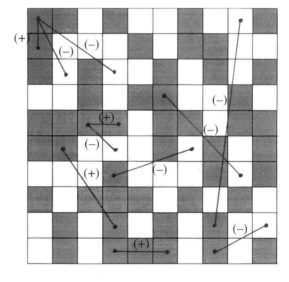

Figure 3.2. A possible memory in the Hopfield model for a network with a hundred neurons. Here a white cell represents a firing neuron (+1 state), and a black (gray) cell represents a quiescent neuron (−1 state). The figure shows how some of the synaptic efficacies (or weights) would be adjusted in this simple model if this memory was to be stored in the network using the special symmetric form of Hebb's rule detailed in the text. In a fully connected network with a hundred neurons there are 4,950 synaptic two-way connections, ignoring self-interactions.

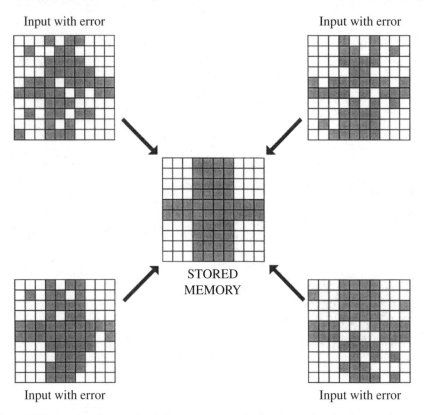

Figure 3.3. Examples of inputs that belong to the same basin of attraction for the stored memory shown. This error-correcting mechanism enables neural networks to recall memories from incomplete or noisy input. In this case, the memory is recalled by the content of the input and not the address of the memory, as in a computer. This phenomenon is referred to as "content-addressable memory."

to the one shown in figure 3.2. In this case, each of these memories uses all of the hundred neurons for its representation and all of the 9,900 synapses for its storage.

If the network is presented with an input pattern that is similar but not identical to one of the stored patterns, it will generally process that input (according to the dynamics described below) until it converges to the "nearest" stored memory state. This is the basis of content-addressable memory in the model. The set of all input states that converge to a particular stored memory is said to belong to the "basin of attraction" for that memory. Figure 3.3 shows examples of four initial states, among many others, that converge to the same stored memory shown.

The dynamics of the Hopfield network proceeds as follows: A neuron is selected at random, and its activation state is upgraded by evaluating the input from all other neurons, each weighted by the appropriate synaptic efficacy for that link. If the sum of all of these contributions is greater than some threshold (usually taken to be zero in the simplest model), that neuron will itself fire, or convert to a +1 state if it is not already in that state. If the sum of the inputs is less than the threshold (here less than zero), the neuron stops firing if it was firing, or remains in the quiescent state (-1). A particularly unrealistic feature of the simple Hopfield model, as noted above, is that neurons are symmetrically connected to each other and act on each other symmetrically. That is, the action of neuron A on neuron B is the same as the action of neuron B on neuron A. As we saw in chapter 2, the flow of information between two neurons in the brain is usually unidirectional and thus is not symmetric. The brain is however, highly recurrent and information indirectly flows back to the original neurons. The Hopfield model has the ultimate level of recurrency.

The procedure of selecting a neuron at random and updating its state of activation is repeated until all of the neurons in the network have been updated. This is referred to as a "network update cycle." In a simple simulation, I stored ten patterns (as in figure 3.2) in a network with a hundred neurons. I then prepared an input that looked like one of the stored memories, but with 30 percent of its activation states switched with respect to that memory. It took just one neural network cycle for this network to recall the stored memory. When I switched the firing activity of 40 percent of the neurons associated with the memory, it took six network update cycles to converge to the stored memory. It is quite remarkable that this network was still able to recall the stored memory, and so quickly, given that the input was so different from the stored memory and there were other memories stored in the network at the same time. The neurons were able to help each other recall the stored memory, even when so many were initially in the incorrect state.

According to Hopfield, the random asynchronous updating procedure is more realistic biologically than a sequential procedure (where the neurons are updated in some definite order, in much the same way that the pistons of a car fire in a definite sequence) or a synchronous procedure (where all of the neurons are updated at the same time). This is supported by the fact that real neurons seem to be updated at random. One could also argue, however, that the synchronous updating procedure is more realistic, because it allows neurons that are ready to fire (that is, have sufficient excitatory input) to be updated without waiting for their turn. Biologically, neurons are updated as the need arises. When a neuron receives sufficient excitatory input, it fires. It does not have to wait for all of the other neurons to be updated, and it can be updated many more times than the other neurons. And yet neither the synchronous nor the asynchronous updating procedures capture what is actually going on in a biological system.

The special quality of the Hopfield model is that the network is highly recurrent. Every neuron connects to every other neuron, and signals are sent backward and forward between neurons in the network until a decision is reached (here an attractor). Recurrent neural networks (with attractors) were considered earlier by others, but it was the Hopfield model that gave the recall process a physical interpretation in terms of an energy minimization process (Amari 1972; Caianiello 1961; Hopfield 1982a; Little 1974; Little and Shaw 1978; Rochester et al. 1956). Hopfield was originally a solid-state physicist, and physicists love to think in terms of energy, as many physical systems evolve to stable minimum-energy configurations. In the Hopfield model every input leads to an attractor state. Although we may not want the system to approach an attractor for every input in a real biological network, it did provide a basis to understand content-addressable memory more rigorously, and this generated an enormous amount of interest (especially among other physicists).

William Little and Gordon Shaw had earlier proposed a model that used a synchronous updating procedure instead. The key difference from the Hopfield model was that sometimes this network would converge to a two-cycle, or a state where the network oscillates between two stationary states, instead of approaching a genuine attractor. Asynchronous dynamics is required for the network to always converge to an attractor state. This was originally considered to be a plus, but we would now prefer a model where sometimes (maybe even more often than not) the dynamics does not converge to an attractor at all. This would be useful to represent a situation where the input is not recognized or is not significant.

In the Hopfield model the dynamics always converges to a fixed-point attractor state, where the activation states of the neurons no longer change. Later we will look at more realistic models in which the network sometimes does not approach an attractor but rather wanders around endlessly, hovers around a particular region in "phase space" (states of the system), or approaches the quiescent state where all neurons are "off." The Hopfiled model converges to either one of the stored memories/patterns or to a spurious state that was not intentionally stored in the network. This convergence usually takes just a few network update cycles. If the network is not too heavily loaded (that is, not too many patterns are stored in it), the convergence to a stored memory is usually faster than the convergence to a spurious memory. I will use this distinction, between these two different types of memory states, to propose an interesting theory about the phenomenon of "déjà vu" in chapter 4.

Hopfield has shown that because of the assumed symmetry in the model there is a generalized form of "energy" that decreases with each iteration until the network arrives at a minimum energy state. This state generally corresponds to one of the stored memories, but there are also stationary states, which were not stored intentionally in the network, that are generated internally by the network itself. (These spurious states, which have been mentioned already, are the subject of chapter 4, where I will suggest that they are essential for creativity and continued unsupervised learning.) Figure 3.4 illustrates the main concepts of the Hopfield model in terms of a simplified one-dimensional

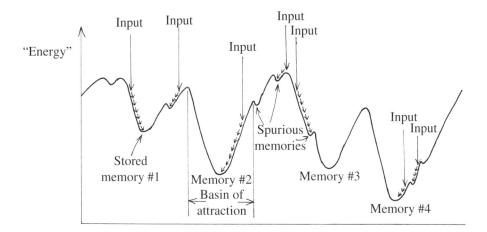

Figure 3.4. A one-dimensional landscape representation of the stored memories in the Hopfield neural network. Here intentionally stored memories correspond to the "energy" minima at the bottom of the large valleys. The smaller local minima correspond to spurious memories that are generated by the network because of the distributed overlapping storage of memory. When the network is presented with an input, it is processed, with the energy decreasing with each iteration until it converges to the nearest local minimum. The situation can be likened to dropping small balls from above and allowing them to roll down into the nearest potential well. The size of the basin of attraction for one of the stored memories is shown. One can see from this diagram how spurious attractors could interfere with the retrieval of stored memories.

representation of network states (Amit 1989). Here the bottoms of the large valleys (or "potential energy" wells, in physics jargon) correspond to stored memories, and the bottoms of the smaller valleys, which are generally located on the sides of the large valleys, correspond to spurious memories. When a network is presented with an input, it processes that input in such a way that, with each iteration, the energy of the system decreases (or remains constant). In terms of the landscape diagram, this means that if an input is presented to the network, as shown in figure 3.4, the state of the network will move down the slope toward the nearest minimum (bottom of a large or small valley) located in its downward path.

As noted earlier, attractor neural networks are accommodating to noise, in that a memory can be recalled from an incomplete input. Figure 3.4 illustrates the range of possible inputs that will converge to the stored memory number 2. An attractor neural network actually works better with a little built-in noise in the dynamics, as long as the network is not heavily loaded. This allows it to jump out of small potential minima, corresponding to spurious memories, and subsequently to converge to the proper stored memories.

SPURIOUS MEMORIES, STORAGE CAPACITY, AND THE OVERLOADING CATASTROPHE

A natural consequence of the distributed and overlapping storage of memory in the Hopfield model (and neural networks in general) is that these networks generated their own set of memories, so-called 'spurious memories', which were not intentionally stored in the network (Hopfield, Feinstein, and Palmer 1983; Amit, Gutfreund, and Sompolinsky 1985). These spurious memories are usually made up of combinations of features of stored memories, but there are also some spurious memories, called "spin-glass states" (after their association with so-called spin-glass models in physics), that are seemingly unrelated to any of the stored memories. One expects that spurious memories also exist in the brain because of the way that real memories are stored in common areas of the brain. A simple example of a spurious memory in the Hopfield model is illustrated in figure 3.5. Notice its close resemblance to each of the stored memories it is made up from.

Spurious memories are generally considered to be a nuisance because they interfere with the retrieval of stored memories. An input close to a stored mem-

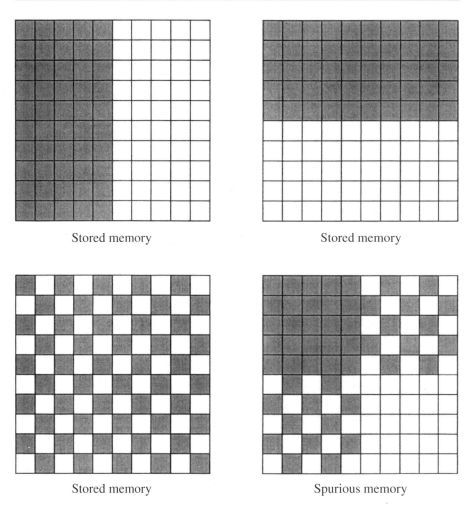

Figure 3.5. An example of a simple spurious memory that is generated in the Hopfield model with a hundred neurons, showing its relationship to the stored memories. This particular spurious memory is derived from the stored memories by a simple majority activation rule for each neuron.

ory may converge to a spurious memory instead of the true stored memory (see figure 3.4). Spurious memories are also a problem because they grow exponentially in number as more and more patterns are stored in the network (Amit 1989). Without any intervention, this uncontrolled growth in the number of spurious memories eventually leads to the so-called "blackout catastrophe," also referred to as an "overloading" or "correlation" catastrophe (Amit, Gutfreund, and Sompolinsky 1985). Generally if more than 0.14 times N patterns are stored in the simple Hopfield network with N neurons, it generates so many spurious memories that it ceases to function as a content-addressable model. Once this storage is exceeded, the network is unable to recall any of the stored memories, even if the input is precisely the same as a particular stored memory. When the number of stored memories is too large and the corresponding number of spurious memories is exponentially larger, the valleys corresponding to stored memory states are swamped by spurious memory minima. A mathematical analysis of the Hopfield model reveals that with excessive storage, the spurious memories start to become the lowest energy states of the system, and eventually the stored memories are no longer minimum-energy configurations.

The overloading catastrophe was originally considered to be a serious problem, particularly because of von Neumann's estimate that the brain acquires about 10^{21} bits of information (one bit corresponds to a binary 0 or 1) in a lifetime, whereas theoretical estimates for a Hopfield network, of the same size and approximate organization as in the brain, could store only a fraction of this amount of information (von Neumann 1958). The brain is thought to be roughly organized into 10^7 networks of 10^4 neurons each (Nauta and Feirtag 1979). A collection of 10^7 Hopfield networks with 10^4 neurons each would be able to store about $0.14 \times 10^4 \times 10^7$ patterns $\cong 10^{10}$ patterns $= 10^{14}$ bits, as each pattern in this particular model has 10^4 bits.

Matters are made somewhat worse when one notes that a considerable portion of the brain (maybe well over one half) is used to process information and not to store it. Incoming visual information, for example, has a lot things done to it (such as edge detection, movement detection, scaling of the size of objects, and contrasting) before it is actually compared with memory. Our estimate of the number of neurons in the brain may also be a little high, as some researchers put the number of neurons in the cerebral cortex at only 10^{10}. The storage capacity of the human brain, considered in this way, may be a low as 10^{11} to 10^{12} bits. Fortunately John von Neumann, father of modern computers and great scientist that he was, overestimated the amount of information that the brain acquires by a few orders of magnitude. He assumed that the brain uses every bit of information passing through the nervous system and that every neuron stores information almost equal to the rate at which it is processing. We now believe that information is stored not by neural processing but by collective stabilized behaviors involving many neurons, such as attractors.

Experiments in psychology also suggest that von Neumann's estimate is much too high. Using results from numerous delayed-memory tests, Thomas Landauer estimates that we store only about 10^9 bits of information in a lifetime, or 100 megabytes (1 byte = 8 bits) (Landauer 1986). He obtained this estimate by asking subjects to learn text and pictures presented to them. Afterward they were tested to see how much they actually acquired in memory. It was considerably less than von Neumann's estimate, but is probably an underestimate because it does not take into account the fact that a considerable portion of information in the brain is generated internally. There is also information contained in the storage of associations between different memories and in temporal memories. Such experiments are also unable to estimate the proportion of irrelevant and weak memories that are stored in the brain, some unconscious and some not easily retrievable. Taking these matters into account could easily bump up the storage capacity of the brain by a couple of orders of magnitude. On the other hand, these estimates assume that everything learned and tested is never forgotten, which may not be entirely true. The estimates arrived at by Landauer may also be higher than normal, as the subjects were concentrating on their task and might be acquiring information at a faster rate than normal.

Based on these estimates from psychology, it would seem that the simple Hopfield model, of the approximate size and organization of the brain, may have sufficient capacity to store what we acquire in a lifetime. In any case, the perceived overloading catastrophe and the explosion in the number of spurious memories as a function of the number of stored memories provoked a massive effort to find solutions to these problems. Note, however, that even if the overloading catastrophe is not a serious problem, the proportion of spurious memories still needs to be controlled in some way, as they generally interfere with the retrieval of stored memories. One approach has been to try to increase the storage capacity of the network by utilizing clever learning algorithms. Unfortunately most of these schemes are not biologically realistic, since they involve highly supervised schemes, or schemes with nonlocal global learning rules (involving the adjustment of practically all of the synaptic efficacies simultaneously).

Some interesting ideas (based on neurobiological features of the brain) did however emerge. Some of these ideas include the fact that synaptic efficacies are bounded and decay with time. More realistic models represent neurons as 0 and 1 states (instead of -1 and $+1$ states) and store memories in a small population of neurons (not in all of the neurons, as in the simple Hopfield model).

These ideas reduce the proportion of spurious memories and increase the storage capacity of these attractor neural networks, while making the general model more realistic neurobiologically. What is most interesting about all of these drastic alterations is that attractors are retained in the neural dynamics. Another way to reduce the proportion of spurious memories (and other weak memories) in these models is to increase the neuron's threshold, or that is, the number of excitatory inputs, required to make a neuron fire. Spurious memories are also destabilized in these models if the condition of symmetric interaction is relaxed. In this case, not all input states lead to an attractor. This allows us to model the situation in which an input is not significant to the network.

Another idea to control the proportion of spurious memories, which actually emerged before the overloading catastrophe was formally identified as a problem, is to allow the neural network to unlearn (Hopfield, Feinstein, and Palmer 1983). Such a process was suggested to be operative during REM sleep when we dream (Crick and Mitchison 1983). This is discussed in more detail in chapter 5.

In the Hopfield model, memories are stored in the entire network and consist of about half of the neurons in the $(+1)$ firing state and half in the (-1) quiescent state. A much more realistic situation is to store memories in only a fraction of firing neurons, say about 10 percent, as in actual neural networks. To model this situation, one has to take the neural states to be either 0 (for quiescent neurons) or 1 (for firing neurons), instead of the traditional ± 1 states. The reason for this is so that the quiescent neurons, which are now more plentiful, do not contribute to the dynamics. Models with a lower "coding level" (the term used to refer to the percentage of active neurons per memory) have a smaller proportion of spurious memories and can store more memories than the Hopfield model. This is because there is considerably less overlap between the stored memories. If memories are stored in too small a number of cells, however they will virtually have no basin of attraction, and there will be no content addressability.

We should note that although the number of patterns that can be stored in models with a lower coding level may be higher than in the standard Hopfield model, the overall storage capacity is still around 10^{14} to 10^{15} bits. The reason for this is that although individual networks or modules may be able to store more patterns (possibly ten times as many), these patterns are generally made up of considerably fewer active neurons (bits).

One of the other outstanding ideas to emerge from the search for a solution to the overloading catastrophe was that the synaptic efficacies should be bounded, or should take on only a limited set of values (Gordon 1987; Parisi 1986a; Peretto 1988, 1992; Nadel et al. 1986). This notion has a neurobiological basis and is now universally accepted by theoreticians, with some experimental support (Bliss and Collingridge 1993). It corresponds to the situation where the release or detection of neurotransmitter is limited and probably quantized in packets of transmission. In models with bounded synaptic efficacies, older memories are effectively discarded as the network learns new memories. Such networks are usually called "palimpsest models." (The word "palimpsest" refers to writing on material that is used again after the first writing has been partially erased, such as when stone parchments were used over and over again in ancient Egypt, with each previous message almost, but not totally, erased.) The storage capacity in the standard ± 1 Hopfield model with a synaptic bound is lower (about 0.04 times N patterns) than in the standard Hopfield model without a synaptic bound (about 0.14 times N patterns), but the former model can continue to function forever without overloading.

In palimpsest models, the synaptic weights are usually allowed to fluctuate between two bounds. When a synaptic efficacy reaches the upper bound, it stays at that level until a new memory is presented to the network to learn, which requires that synaptic efficacy to decrease. If a new memory requires that synaptic efficacy to be increased, the network ignores this. (Vice versa for the synaptic lower bound.) Another way to construct a palimpsest model is to allow the synaptic efficacies to decay with time (Mezard, Nadel, and Toulouse 1986; Nadel et al. 1986). This idea is neurobiologically inspired, as the amount of neurotransmitter stored in the presynaptic vesicles presumably decays with time.

One of the most unappealing features of the Hopfield model (which paradoxically is what got people interested in it in the first place) is that every input converges to an attractor. This undermines the notion that attractors represent significant cognitive events. Sometimes we would like the network not to reach an attractor, so we can say that the input was not significant to the network, or that it was not recognized. If one relaxes the symmetry in the synaptic efficacies, the network does not always approach an attractor. Some (insignificant) inputs will keep changing chaotically and will not settle to an attractor or fixed-point state. Including this biological feature adds psychological realism to the model. It is also interesting that models with asymmetric synaptic couplings tend to destabilize the spurious in preference to the stored memories (Parisi 1986b; Hertz, Grinstein, and Solla 1987; Feigel'man and Ioffe 1987). In asymmetric models, some genuine memory attractors are destabilized as well, but they are generally replaced by pseudoattractors, states in which the system hovers around a particular region of phase space for a prolonged period of time instead of approaching an actual attractor.

In more realistic models, where the proportion of spurious memories is much lower and bounded, one can contemplate the possibility that spurious states are useful (and possibly even essential). Spurious memories may be required for us to generate new ideas and to learn something new (Christos 1995b). (See discussion in chapter 4). Spurious memories also allow neural networks to generalize and categorize memories (Fontanari 1990).

I pointed out earlier that alterations in the neurochemistry of the brain may lead to different memory stores, with some memories belonging exclusively to one particular chemical state of the brain. If certain neurons are excited by a particular drug or neurochemistry, or if the synaptic efficacies are altered in some way, this may change the attractors (and hence the memory) that a network can recall. Attractors in the drugged state may not be attractors in the drug-free state. In terms of the landscape metaphor (figure 3.4), this means the valleys may be shifted or be quite different between these two different states of mind. The same thing may happen when we are dreaming; a different neurochemistry may produce a different set of memories.

THE FORGETFUL BRAIN

An important question that is asked by neuroscientists and psychologists is whether memories are permanent or decay with time and are forgotten. In my view, based on the models considered here and what goes on in the brain, it is inevitable that memories are eroded and lost with the passage of time. (From a personal perspective too, I believe we do forget things. My calculus students forget what I taught them the previous week. Sometimes I find myself rediscovering things [usually things that did not work]. This suggests that attractors for those memories were still there, as I was able to recognize that I had already known about these things. There are other times, however, that I have completely forgotten my computer password or my PINs, and when I was told them, they did not ring a bell.) There is also some evidence from experiments to suggest that animals forget and that synaptic changes induced by learning perish (Squire and Kandel 1999).

As noted before, memories may last a few seconds, a minute, a week, or for years. It is also evident that our recollection of past events changes with time. We find that some memories can never be recalled, no matter how hard we try or what cues we use to try to revive them. Many memories we are having right now may be regarded as quite trivial, and are remembered for only a day or two, such as how many cups of coffee we had today. Such memories will be lost. The question then is, how do memories decay, or how are they forgotten?

It makes biological sense that old and irrelevant memories should decay and be discarded when they are no longer useful. This would give animals a biological advantage in their ability to concentrate on more recent, and potentially more important, memories. The loss of unwanted memories reduces the interference between useful memories and potentially hazardous spurious memories.

Mathematical models like those considered in the previous sections support the notion that the brain does forget. Some attractors (or memories) are converted into states that are no longer stable minimum-energy configurations, and hence disappear. Memories also change with time because they interfere with each other. Old memories determine which new memories will be learned by determining which attractors a network attains, but the storage of these new memories/attractors also changes the synaptic structure in ways that change the old memories. Simulations with simple models demonstrate that memories evolve as the network continues to learn.

One way in which memories can decay is if the amount of neurotransmitter stored in the presynaptic vesicles is limited or diminishes with time. It is unlikely that the brain can maintain its levels of neurotransmitter indefinitely. Another way that memories can decay and be forgotten is if synaptic efficacies are bounded. This causes the network to forget its older memories in preference for newer memories.

Yet another way that memories can decay is if we unlearn them during dream sleep, as suggested by Crick and Mitchison. This mechanism, which was proposed as a means to control spurious memories, will also weaken and eliminate other stored information. Weak memories will be eliminated and older memories will be eroded as they endure more unlearning sessions, or nights of sleep and dreaming. Another way that older and weaker memories could be discarded is if the brain rehearses the strongest memories during sleep. (I will discuss these various options in more detail in chapter 5.)

We noted in chapter 2 that the brain goes through periods of rapid synaptic growth followed by periods of intense pruning, especially during the early stages of development of the nervous system. Such processes may continue, at a subdued level, into adulthood. It is inevitable that some memories will be forgotten when pruning takes place, but also when new connections are made. It also stands to reason that some synaptic connections that are not used very often should be eliminated, as it does not make sense for the brain to waste resources on them. This is another way that learning, and unlearning, can take place in the brain. Memory can be lost when synaptic connections are severed.

From studies of H. M. it appears that the hippocampus slowly offloads its memory to the neocortex over a period extending up to three years. In addition to acting as an addressing and binding system, the hippocampus may act as a temporary store of memory itself. If memories are not encoded with sufficient intensity and repetition, they may not be relayed from the hippocampus to the neocortex, allowing them to be forgotten.

It is interesting to note that the hippocampus is a highly interconnected (recurrent) neural network, even more so than the neocortex (which has a more layered or laminated structure). In this regard the hippocampus forms a structure where attractors may be even more important. In the brain the two hippocampi (remember there are two of almost everything in the brain) contain about 10^8 neurons. Like the neocortex, the hippocampus is made up of smaller modules, called lamellae. If we take the two hippocampi to consist of roughly 10^4 networks of 10^4 neurons each, then their storage capacity is (using a Hopfield net estimate) approximately 10^{11} bits, which is quite large, given Landauer's estimate that we acquire only about 10^9 bits of information in a lifetime. One of the reasons to think that the hippocampus may store its own memory and not just act as an addressing system is that minor damage to it usually affects recent memory. This is especially evident in Alzheimer's disease, where damage to the hippocampus occurs.

FURTHER READING

Memory: From Mind to Molecules, by Larry Squire and Eric Kandel. Scientific American Library. New York: W. H. Freeman and Company, 1999. This is a recent book on memory, written by experts in the field, who explain in considerable detail how memory is stored in the brain and how the various chemicals work together at the synaptic cleft to store memory. Kandel was awarded the Nobel Prize in medicine in 2000 for his work on how memory is stored through modifiable synapses and for his explanation of the molecular mechanisms involved with the storage of memory.

"Neural Networks and Physical Systems with Emergent Collective Computational Abilities," by John Hopfield. *Proceedings of the National Academy of Sciences USA,* vol. 79, pp. 2554–2558, 1982. This is the original paper by Hopfield that started the craze among physicists to study mathematical models of memory. It is quite amazing that many subsequent developments in the field of attractor neural networks, such as temporal sequences, palimpsest models, synaptic dilution, generalization (see chapter 4), and asymmetric networks, were all mentioned in this classic paper. This is one of the most cited papers in research on neural networks.

Modeling Brain Function: The World of Attractor Neural Networks, by Daniel Amit. Cambridge: Cambridge University Press, 1989. Daniel Amit is another of the pioneers and founders of attractor neural networks. This is an excellent book for physicists who wish to become acquainted with how attractor neural networks work. (It inspired me to get interested in the brain.)

CHAPTER 4

The Creative Brain

Unlike a computer or other recording devices, the human brain is extremely creative. It has the capacity to generate memory states that were not intentionally stored in it. I suggest that this ability arises because of the way memories are stored in the brain, in an overlapping fashion, with different memories sharing the same neurons and synaptic connections. Mathematical models show that this type of storage leads to new memory states, which are made up of combinations of features of the stored memories. There is every reason to believe such states should also exist in the brain.

Deeper reflection suggests that internally generated states may also be required for a neural network system to be able to learn new memories, to generalize and categorize, to make associations between memories, to think, and to make mistakes.

COPY OR CREATE

Almost everything we know, do, and believe in as individuals (and as a society) we learned from someone else. Either we had it shown to us, we read it somewhere, we copied someone, or we were influenced by someone. The way we talk, write, read, draw, think, do arithmetic, argue, go about our lives, and even ride a bicycle has all developed and is passed on in this way. We learn things from our family, friends, workmates, teachers, neighbors, and people we meet, as well as from books, newspapers, magazines, scientific journals, television, radio, the Internet, and so on. We thrive on knowledge and the spread of information. There is a very good reason why this is so, based on "memes."

A meme is not precisely defined, but is generally taken to be something, like a skill, a technique, or a useful idea, that we have copied from someone else, such as how to make a fire, how to use tools, how to grow crops, how to read, how to speak, or how to run a business, or just a piece of information or knowledge generally. Memes are supposed to evolve in much the same way that genes do. They meet the three basic requirements for an evolutionary system (Blackmore 1999; Dawkins 1989). Like genes, memes can be copied from person to person (or reproduced in biological jargon); they can be varied during transmission (corresponding to mutation in a biological system); and the best or fittest memes win over other memes (just as the best genes, or expressions of those genes, win over other genes).

Biological evolution has led to some amazing forms of life (such as human beings, elephants, fish, birds, insects, and trees), with an array of unusual and sophisticated biological, biochemical, and biophysical features, such as binary vision, movement, flight, hair, metabolic systems (to produce energy from food), photosynthesis (to produce food from the Sun's energy), and nervous systems, to name just a few. If memes really evolve like genes, what has evolved from their constant refinement? Before answering this question, I note that in addition to evolving, memes have a tendency to form conglomerations of larger memes. As an example, bits and pieces of information are put together to form a scientific theory, or knowledge in various engineering disciplines is used to design a motor vehicle. One could argue that memes are responsible for all of human culture—our laws, politics, religion, science, technology, computers, the Internet, transport systems, and cookbooks, to name a few. All these things, and practically everything else around us that is human-made, have been derived from the evolution and assembly of ideas spread among humans.

Memes offer an explanation for human behavior and the many things we do that do not have a simple biological explanation, such as why we talk so much and freely exchange ideas. Susan Blackmore has gone one step farther, suggesting that the evolution of memes may also explain why humans have language capabilities, why we have such an enormous brain (or neocortex) compared to other animals, and why we believe there is a central "self" with "free will" (Blackmore 1999). It is unusual that we look similar to other mammals (we have many common features, such as two eyes, two ears, four limbs, hair or fur, and bodily organs) yet we are so different mentally. Blackmore argues that we need a big brain to imitate and copy each other, language aids the spread of memes, and the self is a self-supporting collection of memes (or beliefs) that aid their own survival. What makes us special with respect to other animals is our extraordinary ability to copy and imitate each other, which is a profoundly intelligent activity that most animals are not very good at (Blackmore 1999, 2000).

If we just copy each other, though, where do new ideas come from? The (human) brain also has the natural capacity to be creative—that is, to generate something completely new, something not formally acquired. Creative states are generated quite naturally in neural network systems due to the overlapping storage of memory. It could be argued (see, however, comments below) that some new ideas, or creativity, can be generated by the imperfect transmission of memes (or mistakes in communication). I suggest later that this capacity to generate new ideas through the imperfect transmission of memes actually also relies on the existence of creative states in individuals. This is how we make "mistakes."

Creativity is a well-recognized human capability that is highly encouraged, nurtured, envied, and rewarded. If the ability to imitate is an intelligent process, as suggested by Blackmore, then the ability to be creative should be construed as an even more intelligent process, as most humans are not very creative. Some people undoubtedly think that creativity comes out of the blue, or from the "ether" (the mythical fluid or gas supposed to fill all surrounding space), but from a scientific perspective it stands to reason that individual creativity is a function of the brain, and what is stored in it. The question is, how can it arise? I believe the answer is related to the fact that memory is stored distributively in wide areas of the brain in such a way that different memories overlap each other, or use common neurons and synapses for their storage and representation. This overlapping storage of memory in common areas naturally gives rise, as we will see below, to new states or memories that were not intentionally stored in the network. Mathematical models demonstrate that this is a natural consequence of the distributed storage of memory. Novel "spurious memories" usually possess subtle combinations of features of the stored memories, and thus have the capacity to generate new ideas that combine different bits of information. They can lead to lateral associations between memories that are not related in the usual scheme of things. I suggest that these spurious states are the origins of creativity in the human brain.

DISTRIBUTED STORAGE AND SPURIOUS MEMORY

The way memory is stored in the brain is very different from the way it is stored in a conventional computer. This is why the brain is capable of generating new ideas and is creative, while a computer is not.

In a computer, memory, or the value of every variable in the computer's programs, is stored somewhere in the computer's hardware (on its hard drive, or in random-access memory) as a binary string made up of a bunch of 0s and 1s. The value of a variable is retrieved by using the variable's address, which is another binary string that tells a program where exactly it is stored. In a computer, memories are not confused or combined unless arranged so by a programmer.

In the brain, memory is also stored as a string of 1s and 0s, corresponding to whether neurons, labeled in some specific order, are firing or quiescent respectively, but there is no address. Memory is recalled or retrieved from imperfect input, without any specific address. For example, we recall a verse from a song by using some part of it as a cue, and we recognize people we know, even though they (almost certainly) look quite different from when we last saw them. A computer has difficulty retrieving information from imperfect input, and the task of recognizing someone's face from different pictures is a complex problem in conventional computing. This type of memory recall in the brain is referred to as "content addressable," as it is facilitated by the content of the input. In effect the memory is its own address. (See chapter 3 for more details.)

Also, quite unlike a computer, memories are stored in an overlapping fashion over wide areas of the brain, often utilizing some of the same neurons and synapses. In other words, memories in the brain are stored on top of each other. A natural consequence of this overlapping storage is that the brain generates its own sets of memories, called spurious memories. It is fair to say that most researchers regard spurious memories as a nuisance, and much work has been devoted to devising methods to eliminate or control them, to ensure almost perfect recall of stored memory. My contention here is that spurious memories may be extremely useful (and possibly essential), not just for creativity but also so that a neural network can learn something new, adapt, generalize, classify, think, and make new associations.

Because a conventional computer stores memory perfectly with specific addresses, it is generally incapable of generating something new by way of error or by combining information, other than as it is instructed, so it cannot be creative. This however may not be true of all computers, particularly neural computers (also called "neurocomputers"), which are built around a neural network architecture, or a highly interconnected vast network or cluster of computers connected to each other on the Internet.

We saw in chapter 3 how content-addressable memory works for a simple theoretical model called the Hopfield model (Amit 1989; Hopfield 1982a). I would like to use this model here to show how neural networks can easily generate new (possibly creative) information. In the Hopfield model, a special symmetry is imposed on the system that allows for a physical interpretation of the memory-recall process in terms of the minimization of an "energy" function. As we saw in figure 3.4, as long as a network is not heavily loaded, the memory structure can be envisaged as a function that looks like lots of hills and valleys, with stored memories corresponding to the deepest valleys (minimum energy), and smaller valleys located on the sides of the large valleys corresponding to spurious memories. When the network is presented with an input, visualized as a ball dropped from above onto this landscape, the input proceeds (or rolls) down the slopes to the nearest valley/energy minimum, as shown in figure 3.4. In the simple Hopfield model, every input converges to the bottom of a valley. This final resting place could be a stored memory or a spurious memory. In this model, memories, including spurious memories, are represented as fixed-point states, or attractors, in the neural dynamics. Attractors occur when the system stabilizes its activity and reverberates in a constant state of excitation. When the network converges to an attractor, the same neurons fire persistently.

In the standard Hopfield model, the neural activation states are either $+1$ or -1, according to whether neurons are firing or quiescent, respectively, and memories correspond to a certain pattern of activation states. As noted before, a more realistic scenario would be to take the neural activation states to be either 1 or 0, but this does not preserve the symmetry imposed on the model, which allows for an elegant physical interpretation of memory in terms of an energy landscape.

I mentioned in chapter 3 (see figure 3.5) that this simple model generates its own sets of memories, called spurious memories, which generally consist of combinations of features from various other memories stored in the network (Hopfield, Feinstein, and Palmer 1983). These spurious states are generally considered to be a nuisance because there are so many of them. Their number grows exponentially as we store more and more memories in the network (Amit, Gutfreund, and Sompolinski 1985). With more stored memories, there are more and more combinations of memory features possible, and hence more spurious memories. In fact, if we store too many memories in the network, it generates so many spurious memories that none of the stored memories can be recalled, and the network ceases to function as an associative-memory model. This, as we have seen, is referred to as the overloading or blackout catastrophe. Spurious memories also interfere with the general retrieval of stored memory, as is evident from the landscape diagram in figure 3.4. There are however a number of ways to prevent the blackout catastrophe and limit growth in the number of spurious memories, which I will discuss below.

Another problem with spurious memories in simple models, and in particular with interpreting them as creative states, is that they have a lot of symmetry, something we would not readily associate with creativity, which is supposed to involve novelty, certainly not symmetry. The problem here is not that spurious memories are constructed from combinations of stored memories but that there is much symmetry amongst themselves. (See examples and discussion below.) I will argue, however, that this symmetry is broken, and the number of spurious memories significantly reduced or limited, by incorporating some biologically realistic features from real neural networks into the simple Hopfield model.

One may object to the concept of using spurious memories as creative states, made of combinations of features of stored memories, as this would seem to be too restrictive. We will see, however, that there are an enormous number of spurious states and combinations of features that can be constructed in this simple way. The situation is similar to the extensive range of words and phrases we can construct with twenty-six letters, or the endless variety of music we can create with only seven or twelve basic notes. I also suggest that most creative ideas indeed seem to be of this nature, combining different bits and pieces of information together. Nothing comes completely out of the blue, as it were. When we seek a solution to a particular problem, we generally use various ideas, facts, and memories to arrive at a solution that utilizes and satisfies most of the facts and constraints simultaneously.

What is interesting about spurious memories is that they correspond to the shallow and narrow minima on the memory landscape diagram, so individually they are difficult to select or locate. This means that input needs to be close to a particular spurious state for it to be selected, precisely what we would like to

78 MEMORY AND DREAMS

happen if we wanted to interpret spurious states as creative states. They should be relatively difficult to activate.

SPURIOUS MEMORIES IN THE HOPFIELD MODEL

I illustrate here some of the spurious memories that can arise in the standard Hopfield model. Suppose we have stored the memories shown in figure 4.1 in a Hopfield model with four hundred completely and symmetrically interconnected neurons. Here a white cell represents a firing neuron ($+1$), and a gray cell represents a quiescent neuron (-1). The patterns we have used were specially chosen to be orthogonal; in other words, there is no net overlap between any two patterns. This means the given patterns have the property that for any pair of patterns, exactly one half of their respective neurons are in the same state of excitation, either both $+1$ (firing) or both -1 (quiescent), and exactly one half are in opposite states of excitation with respect to each other. Patterns number 1 and 3, for example, have the same states of excitation for all cells on the left half and opposite states of excitation for all cells on the right half. Patterns 1 and 2 have the same states of excitation in the top left quarter and the bottom right quarter and opposite states of excitation in the top right quarter and the bottom left quarter. These symmetric-looking orthogonal patterns were specially chosen to demonstrate the nature of spurious memories and their relationship to the stored memories. The discussion can easily be extended to random nonorthogonal patterns.

Figure 4.2 shows some of the spurious memories this network has generated as a consequence of the distributed overlapping storage of memory. The spurious memories thus generated seem to involve five-by-five blocks of neurons in the same state of excitation, as in the original patterns, so in effect they carry some of the symmetry inherent in the original patterns stored in the network. The spurious memories shown are thus made up from various combinations of features of the stored memories. They are stable states of the system, and if the network is presented with an input that closely resembles one of these patterns, it will converge to that state and not to one of the stored memories.

The spurious memory shown in figure 3.5 results from the three patterns shown there, by a majority rule. That is, for any particular neuron, the simple

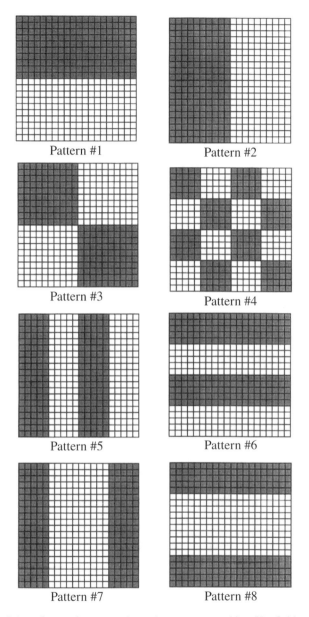

Figure 4.1. The eight orthogonal patterns shown here were stored in a Hopfield model with four hundred neurons. White cells correspond to firing neurons and gray cells to quiescent neurons.

spurious memory shown there takes on the activation state +1 (white) or −1 (gray) if that is the majority value for that site for the three stored memories. For example, the value of the site (or neuron) in the third row and eighth column will be +1 for this spurious memory, because in two of the stored memories, used to construct this spurious memory, the state of this cell is also +1, whereas it is −1 for only one of the stored patterns used.

In general, the minimum number of stored memories that need to be combined to form a stable spurious memory in this model is three, and only odd combinations of stored memories are stable (Amit, Gutfreund, and Sompolinski 1985; Amit 1989).

Figure 4.2 shows some spurious memories that can be generated in the present model by a majority rule using three stored memories. The first spurious memory shown in figure 4.2 combines the stored memories 1, 2, and 3 by a majority rule. The next two spurious memories shown combine the stored memories 3, 5, 6 and 1, 3, 5 respectively. The standard Hopfield model has the special property that the negatives of each of the stored patterns are also attractors or memories. In figure 4.2, one of the spurious memories shown combines the stored memories 1, 3, and the inverse of pattern 2 (denoted by 2*). Each of these 3-mixture spurious memories has 75 percent of its activation states in the same state as each of the stored memories used to create it. As is evident from figures 4.1 and 4.2, they clearly resemble the parent-stored memories they were created from.

Note however that these symmetric 3-mixture spurious states are not direct linear combinations of the stored memories. They involve various features from each of the stored memories, and there is nonlinearity in the way the majority rule is imposed. For the particular model considered here, with only eight stored memories/patterns, there are $2 \times 2 \times 2 \times 56 = 448$ possible 3-mixture spurious states, where the factors of 2 correspond to whether we choose to use the pattern or its negative (one for each stored memory used) and there are fifty-six ways to choose three objects from eight objects.

Stable symmetric 5-mixture spurious states can also be constructed by a majority rule. An example of such a state is shown in figure 4.2, which combines the stored patterns 3, 4, 5, 6, and 8. This 5-mixture state has 68.75 percent ($=11/16$) of its activation states the same as its parent stored memories. It turns out that there are 1,792 possible symmetric 5-mixture spurious states in this particular model with eight stored memories. The number would be much bigger if we had more than eight patterns stored in the network. Then there are symmetric 7-mixture states and so on.

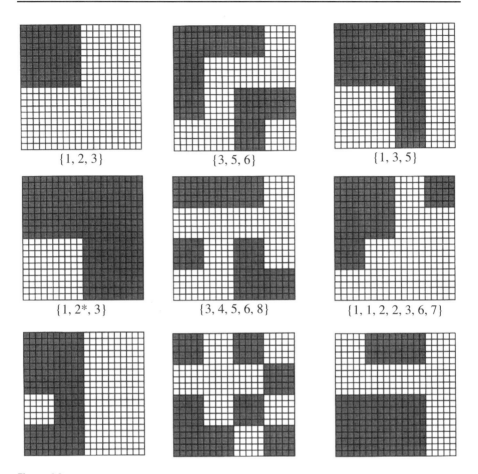

Figure 4.2. Examples of spurious memories generated in the model corresponding to figure 4.1. The spurious memories shown here are made up from specific combinations of the stored memories in figure 4.1. For example, the first pattern is made up from a combination of patterns 1, 2, and 3.

There are many symmetric spurious states. Their number is estimated to be approximately equal to one-half times 3^p (3 raised to the power of p, or 3 multiplied by itself p times), where p is the number of patterns stored in the network. For $p = 8$, as in the example above, this corresponds to approximately 3,200 symmetric spurious memories. For $p = 40$ patterns, there would be approximately 6 billion billion symmetric spurious memories, an enormous number and one of the main reasons why these networks overload when we store too many patterns in them. We can see from this example that even though spurious memories are made up from combinations of stored memories, there are numerous different combinations possible, which gives plenty of scope to generate something novel or creative.

In addition to the symmetric spurious memories, there are also many nonsymmetric stable spurious memories, which combine features from stored memories in unequal proportions. An example of such a spurious memory, shown in figure 4.2, is generated by a similar majority rule but utilizes the stored memories 1 and 2 twice as much compared to the other stored memories 3, 6, and 7. This spurious memory has the same excitation states with 75 percent of the neurons in patterns 1 and 2, and 62.5 percent (=5/8) are the same as those in patterns 3, 6, and 7. Figure 4.2 also shows three other spurious states that were generated in a computer simulation experiment with this particular model. We have not tried to identify what combinations of the stored memories may have been used to generate them.

In addition to spurious memories constructed from stored memories, theoretically there are also so-called spin-glass spurious states, which are reputed to show no resemblance to any of the stored memories. These states are not expected to become important until the network has practically overloaded and are thus not of much interest to us here since we are generally interested in the low-loading level (not too many stored memories per neuron), when the number of spurious states is also low. It is in this regime that I envisage that spurious states may act as creative states. Spin-glass states may, however, have relevance in the early development of the brain, when the loading levels may have been comparatively high as the neural networks were only partially connected and small.

With so many spurious states, it is no wonder that Daniel Amit concluded, "Our attitude here is that spurious states are spurious!" (Amit 1989, 199). Amit was also disturbed by the fact that there is an inherent symmetry, in that all symmetric combinations of mixture states are stable. That is, all combinations of three stored memories and all combinations of five stored memories generate stable spurious attractors, which theoretically are equally important. This seems to be too much symmetry if we are going to associate spurious memories with creativity.

CONVERTING SPURIOUS STATES INTO CREATIVE STATES

In order to make the association of spurious memories with creativity, we need to find a way to significantly reduce their number and break the strong symmetries relating them to each other.

One of the reasons why there are so many spurious memories generated in this model is that there is a lot of symmetry in the model. All the neurons are connected to each other, and their coupling or influence on each other is mutual—that is, the effect of neuron A on neuron B is the same as the effect of neuron B on neuron A. This symmetry is of course not present in the brain, as most neural connections, as we have seen, are unidirectional.

The number of spurious memories can be controlled by placing a bound on the synaptic efficacies. Recall from chapter 3 that when we store a memory in a neural network, we update the synaptic links that connect neurons in the same state of excitation (see figure 3.2.) When we store more memories in the same network, we increase the corresponding synaptic efficacies by one unit. A particular synaptic connection may be increased a few times if it is important to more than one stored memory. The problems with spurious memories emerge when the individual synaptic values become too large. This is when the various stored memories start to interfere with each other. A possible way around the excessive buildup of spurious memories and the overloading catastrophe is to place a bound on the synaptic efficacies. This is a very reasonable thing to do. The efficacy at a synaptic junction is largely controlled by the amount of neurotransmitter that crosses the gap. As we saw earlier, not only is there a finite amount of neurotransmitter available for release into the synaptic cleft, but each presynaptic vesicle releases a specific amount of neurotransmitter when it opens. This means that the allowed synaptic efficacies are not only bounded, but are quantized as well.

In models with a synaptic bound, the synaptic efficacies are increased if the storage of a certain pattern requires them to be increased, but if a prescribed limit is reached for any synaptic connection, that synaptic efficacy is no longer increased. It remains at that limit until it is presented with a pattern that requires it to be decreased again (Parisi 1986a; Nadel et al. 1986). An interesting feature of these models is that they can generally recall only the last few memories stored in them. In a symmetric Hopfield model with four hundred neurons and a synaptic bound, one can store approximately fifteen patterns, compared to fifty in the standard model without a synaptic bound. Of course, the latter model eventually overloads and ceases to function as a content-addressable memory system. On the other hand, the model with a synaptic bound continues to function forever, although it keeps track of only the last few patterns.

Another biological aspect of real neural networks that helps reduce the proportion of spurious memories is that real memories are usually represented with many fewer firing neurons. Recall from chapters 2 and 3 that the human brain can be thought of as consisting of 10 million modules (or densely connected neural networks) with about ten thousand neurons each. A typical memory may utilize around a hundred to a thousand active neurons per module and involve around ten modules. Each activated module may correspond to the various features that make up a complete memory. In the standard Hopfield model, memories are encoded using 50 percent of all neurons in a module as active (in all modules), whereas real neural networks probably use a "coding level" (the number of active neurons in proportion to the total number of neurons in a module) of around 1 to 10 percent. The effect of reducing the coding level, and the number of active modules, is to reduce the overlap between memories and hence the number of spurious memories generated by the network.

Sparsely coded networks can store many more patterns than the standard Hopfield model. Theoretically, the standard Hopfield model with four hundred neurons can store around fifty memories, whereas a neural network with a coding level of 1 to 10 percent can store around a thousand memories (Tsodyks 1989). (This number is larger than the number of neurons in the network, so more memories can be stored in this model than in the old grandmother cell theory, where a single memory is stored in each neuron.) There are many fewer spurious memories in models with a lower coding level with the same number of stored memories. The large number of memories that can be stored in such models is a reflection of this. The price paid for using a lower coding level is that memories have smaller basins of attraction and the network is less capable of recalling memories from noisy input.

In models with low coding levels, as we have seen, it is more appropriate to use the neural activation states 1 and 0, instead of $+1$ and -1, as in the standard model. The reason for this is that in the ± 1 model, -1 represents a quiescent neuron, and quiescent neurons are not supposed to have any effect on the other neurons. In the ± 1 model with a low coding level, however, quiescent neurons (which are in the majority) can produce a huge effect, overwhelming signals from the active neurons. The use of a 0/1 model eliminates this problem, and reduces the interaction or correlation between stored memories, as now quiescent neurons are not bound together in storing memories.

Noise in neural systems helps reduce the effect of spurious memories. A little noise allows the network to "jump out" of some of the small energy minima of figure 3.4, which correspond to spurious memories, and to roll into deeper minima, corresponding to stored memories. Another way to reduce the effectiveness of spurious memories is to increase neuronal thresholds—that is, the number of (firing) presynaptic (input) neurons required to make a postsynaptic neuron fire. In the standard Hopfield model, there is generally no threshold, and a neuron fires if the balance of input is positive. In real neural networks, around a hundred to a thousand input cells must fire concurrently to make a postsynaptic cell fire.

Real neural networks are randomly, partially, and asymmetrically connected. Giorgio Parisi has shown that relaxing the symmetric action of neurons on each other (as in a real biological system) has the effect of preferentially destabilizing spurious memories over stored memories (Parisi 1986b). In asymmetric models, many of the spurious attractors are replaced by "strange attractors" whereas the stored memories remain as genuine attractors.

Yet another way to limit the number of spurious states is to allow the network to unlearn, by which I mean the opposite or reverse process of learning. Instead of strengthening the connections between neurons firing in synchronization, one weakens them. As noted above, Crick and Mitchison have suggested that such a process may take place during dream sleep (Crick and Mitchison 1983). According to them, this unlearning process has the effect of mainly removing spurious memories, as they are more readily excited during dream sleep. (See chapter 5 for more details.)

The symmetry between spurious memories and the number of spurious memories are reduced if memories are not stored with equal intensity, which appears to be something that occurs in reality. Some memories are stored with great intensity, say if they mean something to us (if they heighten our conscious awareness or involve our emotions in some way), whereas other memories are stored very weakly.

Another important property of memory storage in the brain that helps reduce the symmetry of spurious states is that memories are stored in semi-specific locations, depending on their nature. Visual memories are stored in the visual and associative areas, for example. This means that spurious memories are less likely to develop between memories that are not associated with each other or do not use common storage areas. This tends to make spurious memories more meaningful, as they combine related memories.

Because spurious memories have historically been regarded as a nuisance, there are no formal studies of them in more realistic neural network models. And although I am suggesting here that spurious memories may give rise to creativity and prove useful for other brain functions, they still need to be controlled in some way, to avoid potential overloading and problems with the recall of stored memory.

THE NATURE OF CREATIVE IDEAS

We generally solve problems and generate new ideas by combining various notions, ideas, or memories in our mind (that is, bringing them into consciousness), and allowing the brain to generate new attractors by using these attractors as input. We can then "test" these new attractors (which must be spurious attractors to be new) with respect to the problem we want to solve and other stored memory and knowledge. When something "special" has been derived that fits in with everything, we may experience a sudden surge of electrical activity as many neural networks simultaneously light up each other. This is the moment (eureka!) when we know that a solution has been found. A creative spurious attractor suddenly presents us with a solution to our problem that fits in with all of the other constraints. (Note that the activation level of a neural network can suddenly increase beyond normal levels because if more neurons start firing this can induce even more to fire in turn.) These creative solutions are generally comprised of combinations of features of the stored memories used as cues in the thinking process and the memories they themselves may have generated. The solutions cannot be anything else. The old physics principle of "causality" tells us that something does not just happen without a reason or an origin.

The famous mathematician Henri Poincaré (who discovered chaos theory some seventy years before everyone else and invented topology) was intrigued about creative processes in general. His writings show he identifies that creativity is generated within one's own brain in a strange way that seems to combine stored memories and knowledge together: "For fifteen days I strove to prove that there could not be any functions like those I have since called Fuchsian functions. I was then very ignorant; every day I seated myself at my work table, stayed an hour or two, tried a great number of combinations and reached no results. One evening, contrary to my custom, I drank black coffee and could not sleep. Ideas rose in crowds; I felt them collide until pairs interlocked, so to speak, making a stable combination. By the next morning I had established the existence of a class of Fuchsian functions, those which came from the hypergeometric series; I had only to write out the results, which took but a few hours" (Poincaré 1982, 387). Poincaré's comments about pairs of ideas interlocking into stable combinations is especially interesting because this is precisely what spurious memories are: stable combinations of features of stored memories.

We often imagine that creativity is totally new and original, but in most cases it is not. It generally possesses features of known facts (or stored memories). Ideas are built on other ideas and knowledge, and no person is truly and absolutely original. Creative ideas are copied, borrowed, and manipulated versions of what we know and acquire. Human creativity and imitation (or memes) play complementary roles in the evolution of ideas. Creative ideas are required to share with others, to drive the copying system, but on the other hand most of our creative ideas are also influenced by what we see around us, or what we have learned from others. All creative artists and scientists are invariably influenced by others, and although their ideas are often referred to as original, one can usually finds traits from other artists and scientists in their work. Our ideas come from what we know, which mostly comes from others. What we do is combine this knowledge in unusual ways to generate something new. My own theory (see chapter 6) about sudden infant death syndrome (SIDS) results from combining various ideas about memory, dreaming, lucid dreaming, "water babies," and SIDS itself.

The other important point to keep in mind is that if something were completely new and had no relation to anything before it, we would be unable to assess its importance. It is the overlap with other memory states that enables us to ascertain if a spurious memory is useful.

Another way that creativity (or new ideas) seems to arise is when we are stumbling around making (small) mistakes, or adapting to the situation at hand in some way. This can happen if we are not attentive and the input is not fully appreciated, or if the situation is fairly new to us. The human brain is not perfect and makes mistakes, but how do these errors come about? Content addressability works in the opposite direction as it recalls a memory from an imperfect input with errors or noise. We want a situation where the network makes a mistake and converges to the wrong memory state. But if this is just another stored memory, then all it achieves is an inappropriate association, and it is not creative. Ideally we want to generate memory states that are different, maybe only slightly different from the stored memories. If these states are too similar to one of the stored memories, they are likely to belong to that memory's basin of attraction. The only states that seem to satisfy these requirements are the spurious states, which, as noted, can have large overlaps with the stored memories but have their own basins of attraction. Figure 4.2 shows a spurious state (first pattern in the last row) that looks very similar to one of the stored memories (pattern number two in figure 4.1) but has its own basin of attraction.

Another way that creativity may arise is through the imperfect transmission of memes, but this seems to be identical to the previous scenario as it calls for the receiver (or the transmitter) of information to make a "mistake," calling for spurious memories again. Memes can help spread ideas but cannot themselves lead to something new. The human touch is essential to do that, to allow mutations or errors in transmission. However, another possible source of variation in memetic evolution may come from the fact that different people with different memory stores (that is, different personalities) may see the same situation differently, and only parts of certain information may be transmitted.

TAPPING INTO CREATIVITY

If spurious memories are indeed the source of creativity, how can we tap into them? Crick and Mitchison, as we have seen, have suggested that dreaming may be involved with an unlearning process that predominantly affects (and hence stimulates) spurious memories. The excitation of spurious memories during dream sleep is enhanced by the fact that the brain (or neocortex) is stimulated by semirandom signals generated in the brain stem (instead of the usual structured sensory input it receives when we are awake) and is in a state of weakened inhibition. (See chapter 5 for more details.) This may explain why our dreams are so bizarre. If dreaming is mainly associated with the excitation of spurious memories, it may be a good source of creativity. Many artists and scientists try to utilize their dreams as a source of new ideas. (I know of one prominent mathematical physicist who keeps a pad next to his bed just in case he awakes from his dreams with a good idea.) It is also well known that a number of major discoveries have been made during dreaming (or daydreaming). Otto Loewi conceived how neurotransmitters work while dreaming over the course of two nights, Dmitry Mendeleyev awoke from a dream and wrote down the periodic table of the chemical elements, and Friedrich Kekulé thought of benzene rings during a daydream.

The problem with using dreams as a source of ideas is that dream content is usually forgotten as soon as we awake. The reason for this is thought to be that a different neurotransmitter system kicks in when we awake, or that the brain has suddenly switched from an unlearning to a learning mode. If one wants to recall the content of a dream, it has to be done almost immediately, and it seems to be necessary to rehearse backwards what happened in the dream as soon as we awake and write down our thoughts on a piece of paper. Another interesting phenomenon that may be helpful in recalling dreams, and possibly initiating specific dreams, is so-called conscious or lucid dreaming, which is discussed in chapter 5. Lucid dreamers become conscious during the course of a dream. Lucid dreamers seem to be able to remember more about their dreams. It is known that people can become adept at lucid dreaming with a little practice, and there are electrical devices that can boost the prospects of having a lucid dream. One should note, however, that most spurious memories are probably just spurious, and so care should be taken not to place too much emphasis on them all as being creative or meaningful.

Some people, especially artists, poets, songwriters, and composers, tap into creativity through the use of mind-altering drugs, like marijuana, LSD, heroin, opium, alcohol, and even caffeine (as noted by Poincaré). Numerous songwriters and rock stars, particularly during the 1960s, are known to have used drugs extensively (especially marijuana and psychedelic drugs like LSD) to generate new creative ideas. The problem is that a lot of them, like Jim Morrison (of the Doors), Jimi Hendrix, and Janice Joplin, died prematurely from an overdose. Samuel Coleridge wrote his famous poem "Kubla Khan" based on a dream he had while intoxicated with opium. Edgar Allan Poe drank alcohol excessively. The Australian artist Brett Whitley was known to use drugs for inspiration with his painting, before he also died of a heroin overdose. Drugs are obviously not a practical tool for scientists to use to tap into creativity since their work generally requires a considerable amount of knowledge and logical deduction. Scientific ideas need to be assessed with respect to previous ideas and knowledge and tested through experimentation.

Drugs may stimulate creative modes by adding noise to the system or by altering the neurochemistry of the brain. This can subsequently influence the dynamics, the memory stored in the synapses, and the types of memories that can be recalled. Some of the memory states generated in this way may not exist in the unaffected state; hence the remark that drugs "open up your mind." As a corollary, this may explain how many drugs alter our perceptions and lead to a state of confusion.

Both drugs and dreaming, which alter the neurochemistry of the brain, may help retrieve spurious states, which are difficult to reach when we are processing normally. A common feature of these altered states of mind, though, is that often good ideas generated in this way turn out to be silly ideas when tested back in "reality."

Sometimes new ideas come to us when we are not (consciously) thinking about them, or if we change our usual work practices or surroundings. A change of environment could help to stimulate different attractors. Scientists have often reported that they have good ideas when they are thinking least about their work or when they are traveling. Many people claim to have great ideas in the shower. Chaining oneself to a desk is not a practical way to generate new ideas. This is one of the reasons (other than the need to collaborate with others) that many scientists like to travel to exotic places to attend conferences, and often like to climb mountains, for example, while they are there. They do it for inspiration (or new ideas). Poincaré claims he had many of his best ideas while traveling: "I entered an omnibus to go to some place or other. At that moment when I put my foot on the step the idea came to me, without anything in my former thoughts seeming to have paved the way for it" (Poincaré 1982, 387). Poincaré's subconscious was obviously thinking about the problem he solved on this day. He asserts that new ideas are often generated through a process whereby a problem reverts from the conscious mind (required to initiate the "search") to the unconscious mind (required to generate new ideas) and back to the conscious mind (required to ascertain the usefulness of the ideas generated by the unconscious mind).

I believe that some people are born to be creative, or become so during their childhood development, and that dreams, drugs, and excursions are probably not that useful as a means to generate creativity in people who are inherently not creative. Although there are examples of creative ideas originating from dreams or drug use, there are many more examples of great ideas that have not relied on them. It may also be that some creative people use drugs not to be more creative, but to escape from their overly creative reality. It is important to note that it takes a certain level of intelligence or knowledge to appreciate a good idea, but sometimes others (or the masses in general) make that decision for you.

KNOWLEDGE VERSUS CREATIVITY

Albert Einstein once said, "Imagination is more important than knowledge," but if I am correct in asserting that creativity is based on spurious attractors, which combine features of stored memories, then one should supplement Einstein's statement with the proviso that imagination is primarily based on knowledge. Knowledge is also important to ascertain whether a creative idea is useful. One thus expects that "experts" in various fields would be best able to come up with creative solutions to challenging problems in their fields. You would hardly expect someone off the street to come up with a "grand unified theory" or a "Theory of Everything." As the Nobel prize-winning Russian physicist Lev Landau put it in a letter to a crackpot: "Dear Comrade X, ... Modern physics is a complicated and difficult science, and in order to accomplish anything in it, it is necessary to know very much. Knowledge is all the more needed in order to advance any new ideas. It is obvious from your letter that your knowledge of physics is very limited. What you call new ideas is simply prattle of an ill-educated person; it is as if someone who never saw an electric machine before were to come to you and advance new ideas on this subject. If you are seriously interested in physics, first take time to study this science. After some time you yourself will see how ridiculous is the nonsense that came out of your typewriter" (Livanova 1980). Sometimes, though, a real breakthrough in an area of science comes from someone who has just entered the field, who may be conversant with the subject but is certainly not an expert. This may be because they are more willing to consider creative ideas or exotic alternative theories that a specialist in the field might reject almost immediately.

Knowledge is not the only criterion for creativity, as some people, such as those with autism (previously referred to as "idiot savants") can have vast knowledge of facts but little or no imagination and be generally unable to generalize or adapt to new situations. In the movie *Rain Man,* the character Raymond Babitt (played by Dustin Hoffman) recites to a waitress addresses and phone numbers that he had read earlier in a telephone book, and he remembers the cards that had been discarded in a game of blackjack at the casino. Babitt is however unable to cope with social matters, and he becomes upset when something interferes with his usual routine. He is unable to adapt. I will suggest below that the ability to generalize and adapt may also be related to spurious memories, so this would seem to imply that autistic savants have low levels of spurious attractors.

Autistic savants are generally also self-centered; they lack social awareness and the ability to judge others by way of their facial expressions. They are unable to imagine what it is like to be someone else, so they lack self-awareness and self-control. If the lack of spurious memories is the key to their unique behavior, this may also suggest a possible link between spurious memories and self-consciousness.

There are many other people, not just autistic savants, who have good memory (or knowledge) but are not very creative, like some teachers and librarians (only some). (Teachers and librarians invariably appear on and win most quiz shows on television.) There are also people who are reputed to be very creative but have bad memory (like some artists), people who have good memory and are very creative (like some scientists, but not all), and many more people with bad memory and little creativity. The situation is obviously much more complicated than this, and there is a spectrum of people in most professions who fit into each category. What may be true in general is that people with poor knowledge are less likely to be creative because it takes a certain amount of knowledge to be able to determine if a particular creative spurious mode is useful or not. Intelligence, which clearly encompasses something more than just knowledge, may be defined, along with creativity, as the ability to generate and recognize the usefulness or "symmetry" in particular spurious patterns. Intelligence also involves our abilities to generalize, acquire, and adapt, qualities that I will suggest later are related to spurious memories.

How can we explain this apparently paradoxical array of associations between knowledge and creativity? The models we have considered support various alternatives. Generally, as we store more memories in a neural network, the proportion of spurious attractors goes up, as there is more interference, but this in turn forces the number of recallable memories to go down again. (The process is nonlinear.) In this model alone, we have people with poor memory and poor creativity, people with better memory and more creativity, and people with poor memory and much creativity. The situation is also dependent on the coding level, and different brains (or different parts of the brain) may utilize different coding levels. If a neural network has a low coding level, it may be possible to have more stored memories (with smaller basins of attraction) and fewer spurious memories, compared to a neural network with a high coding level. Autistic savants may belong to this category. Alternatively, one can have high coding levels, with more interference between memories and higher levels of creativity. On top of all of this, the neurochemistry is also important here, as it can influence the activation levels in the brain (and hence the coding levels) and the actual memories and types of memories that can be recalled. A certain adjustment in the neurochemistry of the brain can increase the level of noise, which can reduce the proportion of spurious memory if the disturbance is not too high. In this way the system jitters out of the small spurious minima. A larger adjustment in the neurochemistry of the brain, however, may result in larger amounts of noise, which can then excite more spurious memories by drastically changing the dynamics of the system. This could result in more creativity, but too much stimulation may result in confusion.

What would happen if the level of spurious memories became too high? One might suggest that this could be associated with schizophrenia (and false associations) or other forms of madness, or with dementia and Alzhiemer's disease (and the inability to recall recently stored memories). Schizophrenia is characterized by false associations between memories, or disjointed personalities. Spurious memories fit the bill quite well, as they correspond to associations between memories. In the case of Alzheimer's disease, spurious memories can interfere with the retrieval of stored memories.

It is often said that there is a fine line between genius and madness, and spurious memories provide a nice connection between these two states of mind.

ON THE EVOLUTION OF IDEAS

Once creative ideas arise, they can evolve quickly through interactions among people and through their spread in the community—that is, via the exchange of memes. Someone discovers something, which is picked up by others, who can now use this newfound idea or information to generate other ideas and knowledge, which advance on what has been found. For example, someone invents a unicycle (after someone has already invented the wheel), which is incidentally quite unstable. Someone else adds an extra wheel to this system to make it more stable and invents the penny-farthing, which has one big wheel and one small wheel. The sizes of the wheels are then varied until someone discovers that the now traditional bicycle with two almost identical sized wheels is best. As the bicycle continues to be refined, someone else decides to go to four wheels (modeled on the traditional horse-drawn wagon), and someone else adds an engine to it. The engine evolves separately from a steam engine to a petrol engine to a fuel-injected engine. The modern motorcar is born and is then refined with all sorts of added features along the way, such as disc brakes, power steering, steel radial tires, and air bags; and on another tangent, trucks, buses, and airplanes are developed. There is no question that the sharing of information among humans has enabled them to accelerate the process of invention. What is probably happening here is that old ideas are being combined into new creative ideas, or solutions, using spurious states. This can happen if an inventor comes up with a novel creative spurious state or stumbles onto an invention by mistake as it were, which also involves spurious states.

We noted previously that mistakes, or errors in the transmission of memes, can give rise to new creative states, but these also seem to require individual spurious memories. Such spurious memories may be generated if something was not explained properly or if it was misinterpreted. It is important to note, however, that different people have different memory stores and hence may see something differently from others, even if it is transmitted without error. If these modified interpretations lead to better ideas, they may be taken up by others. The evolutionary process refines ideas in this way.

Residing in an information system (be it the scientific community, researchers in a particular field, various professional specialists, or all of human society) is a common understanding of what is useful or not, which drives the system. Good ideas are identified by individuals and by the collective majority, which then encourages others to copy and modify those ideas further. This collective human intelligence (or in some cases collective stupidity) is essentially the fitness function for the evolution of memes and ideas, determining which memes will survive and prosper and which will wither and die. A group of people can in some ways also act like a neural system. Collectively they can correct misconceptions that individuals may have, in much the same way that neurons collectively correct the firing activity of other neurons that are in the incorrect firing mode for a particular stored memory. They can reject or support and propagate a new idea. This begs the question of whether spurious states can also exist in such a system.

This discussion shows how human interaction can accelerate the problem-solving process and result in creative variations. Such thoughts suggest the need to discuss our ideas with others and to listen to what others have to say, but it does not matter too much if we are not completely attentive or if we sometimes misinterpret what they are saying, as this can be helpful in generating new ideas.

SPURIOUS MEMORIES ARE PARTICULARLY USEFUL!

I will suggest below that in addition to providing a basis for creativity, spurious memories may also be useful for learning, thinking, adaptation, generalization, association, and temporal memory.

Hebb's rule asserts that learning takes place when pairs of neurons fire in coincidence, such as when the brain reaches an attractor. If we imagine that our brain is an attractor neural network that receives input and processes it until it reaches an attractor and then learns the resultant attractor, one is confronted with a dilemma. If there were no spurious memories, this network would recall and relearn only what it already knows, because it would converge only to its stored memories. Without spurious memories, the brain would not be capable of learning anything new, and furthermore it would become obsessed with its own strongest stored memories. Spurious memories help the brain avoid this problem and learn something new, albeit similar in some respects to what is already stored in the network. (But as we saw earlier, the variety of spurious memories is practically endless.) When the network is presented with something new that does not correspond to a stored memory, if it has any relevance to the neural network it may converge to a spurious attractor, which is of course related to the stored memories. If an input is completely "uninteresting" (to the brain), no attractor at all will be achieved. If the network converges to a spurious attractor, it can then sculpt out this spurious attractor by enhancing its corresponding synaptic connections so that it is now an important stored memory.

Figure 4.3 illustrates the possible evolution of a neural memory landscape as it learns new spurious memories and strengthens existing memories. As this process proceeds, there may be other processes at work, such as the decay of synaptic efficacies (or degradation of stored neurotransmitter), the dynamics of bounded synaptic efficacies, and reverse learning, which weaken the stored memories.

There is a slight problem with the idea that spurious memories are required to learn new memories. If autistic savants are not very creative, they probably have fewer spurious memories, but they are nonetheless able to learn more facts. A possible way out of this dilemma would have autistic savants utilizing a different class of spurious memories, such as the spin-glass spurious states, which have no resemblance to stored memories, so the usefulness of these states would be difficult to ascertain. Alternatively they may utilize the same types of spurious states as others but for some reason be unable to ascertain their usefulness. It has been suggested that autistic savants learn everything that comes their way because they lack the ability to realize that some things are unimportant and can be ignored. Normal humans discard large amounts of data that come their way, and this seems to be important to function properly in a complex society.

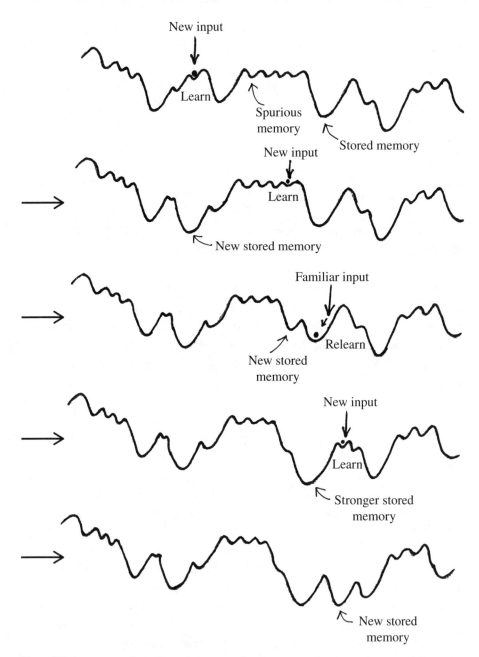

Figure 4.3. A representation of how the memory landscape may develop as the network learns its own attractors, corresponding to the relearning of stored memories and new learning of spurious memories.

If new memories are formed by the strengthening of spurious memories, one may ask how it is that the brain got started in the first place, or in other words, how the first few memories were stored in the brain. One possibility has to do with the way the brain develops and learns during childhood, making many random connections, which are later pruned as we learn (Changeux 1986) (see also chapter 2). This is particularly interesting because initially the randomly connected brain could be viewed as a system that has only a few intentionally stored memories, so the only spurious states available are the spin-glass states. As the brain learns, it starts to select from these shallow spin-glass states to form its first few new memories (Toulouse, Dehaene, and Changeux 1986). As the memory structure takes form and the growth and pruning process subsides, the stored memories take over, and learning can then proceed by selecting from the more conventional spurious memories made from combinations of features of the stored memories.

Another problem with the suggestion that spurious memories are required for learning new information is that the human brain has the uncanny ability to distinguish between recognizable stored memories and new spurious memories to be learned. We know when something is familiar and when it is not, so if spurious attractors are given the status suggested here, how does the brain distinguish between these two different attractors? A possible solution to this may be provided by the fact that in mathematical neural network models the convergence rate to stronger stored memories is much faster than the convergence rate to spurious memories, so the brain could distinguish between these two types of attractors in this way.

This is particularly interesting because it may offer an explanation of the previously inexplicable phenomena of déjà vu (or the feeling that a situation is familiar, or has happened before, when it should not be familiar, or has never happened before) and jamais vu (when something is unfamiliar when it should be familiar). Déjà vu may correspond to the situation where for some reason the convergence to a spurious attractor is rapid, and the brain confuses this spurious state with a stored memory. Jamais vu, on the other hand, may correspond to the situation where the convergence to a stored memory is much slower than it should be, and the brain does not recognize it.

The adaptability of neural networks may also be linked to spurious memories as they allow a network to make small mistakes and use these newfound states to adapt to a novel situation. Adaptation is not possible in a perfect recording device like a computer. However, as noted earlier, neural networks generally accept and function with errors. That is how content addressability works. What we want here is for the neural network to make errors. This can be achieved by using spurious memories, which are similar to stored memories. This idea is interesting because when we adapt to a new situation, we generally combine various pieces of knowledge together to arrive at a new way to deal with changes to our environment.

Spurious memories may also be important to make (new) associations between different memories and ideas. When we learn something new, we invariably have to place it in context with what we already know, which means that we are forming associations with other stored memories. Since spurious memories are already made up of combinations of features of stored memories, they can naturally incorporate such associations, and this may be how the brain in fact develops associations and categorizations. For example, the emergence of the spurious attractor combining patterns 1, 2, and 3 in figure 4.2 may be how the brain makes an association between these memories, and how the brain places this new stored memory (corresponding to that spurious memory) in context with other stored memories.

Jose Fontanari suggests that spurious memories may also play a crucial role in the ability of Hopfield attractor neural networks to generalize (Fontanari 1990). Generalization is the ability of a neural network (or the brain) to learn a concept from a limited supply of examples that typify that concept. Say, for example, we are shown a collection of animals and told they are all "dogs," and then we are shown a new animal that we have never seen before, which also happens to be a dog. If we have been shown a sufficient number of exemplary examples, we should be able to generalize and make the association that this animal is also a dog, or belongs to the same category as the other dogs we were shown. Based on what I have been saying above, the only way a neural network can learn new memories and come up with associations between new memories and previously stored memories is via spurious memories, so we should not be too surprised that spurious memories may play an important role in allowing the network to generalize. Fontanari and Theumann put it quite elegantly when they say, "Storing correlated patterns in the Hopfield model is a problem that involves mixture states since the network's state must have either a non-zero overlap with all the patterns or with none" (Fontanari and Theumann 1990).

Fontanari also suggests that the ability of the Hopfield network to generalize and categorize (that is, maintain and generate a number of different concepts) is spontaneous (Fontanari 1990). As long as the network is presented with a sufficient number of exemplary examples of a concept that are not too different from each other, it can generalize to that concept, and to a number of such concepts. What may be happening here is that by storing many similar patterns in the network, the network generates lots of spurious states, which are very similar to these stored memories. They then start to develop into a larger single stored memory that encompasses all of the examples of that class or concept, and more.

Figure 4.4 shows a diagrammatic representation of how this ability to generalize may come about. If we home in on a particular memory and consecutively store more and more patterns in the network similar to this memory, this generates an array of related spurious memories, which will start to merge into one larger single memory representing that concept. The Hopfield network is able to learn many such concepts or pattern intraclusters in this way, and as long as it does not learn too many concepts, the number of spurious intercluster solutions will not get out of hand. The ability of the network to generalize and categorize is the natural way that these neural networks seem to handle the situation where they are presented with large amounts of (correlated) data.

If my notions about the learning of spurious memories are correct, the new memories, which are subsequently stored in the network, are just spurious attractors that already existed in the network. In this case, as we store more similar memories in the network, the corresponding set of spurious memories is being used up in generating the concept memories, and this may explain why we are able to store many concepts within the same network. Incidentally, the generation of concepts tends to smudge the finer differences between the previously acquired distinct memories belonging to the same concept. This process enables the brain to forget about how to distinguish between similar memories. This is a general phenomenon of human memory, and provides another mechanism for how neural networks can forget through interference.

In chapter 3, it was suggested that one way in which the brain may develop dynamic or temporal memory is by utilizing a sequence of attractors, one leading onto the next. One way to achieve this in neural network models is by introducing time-delay synaptic couplings, which explicitly connect consecutive static memories together. The dynamics is generally constructed so that a particular attractor is achieved using fast synaptic connections. This attractor is subsequently destabilized by slower synaptic connections, which drive the system into the next attractor in the sequence a short time later. In practice the

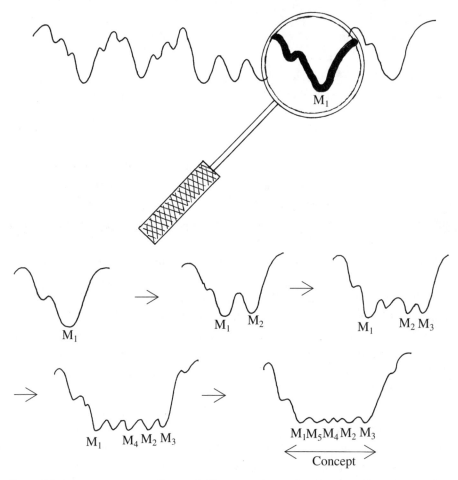

Figure 4.4. Attractors corresponding to similar memories tend to fuse together to form larger, broader-based memories. This is how a network develops the ability to generalize.

snapshots or attractors that make up a dynamic memory are very similar to each other and so are highly correlated. This is where spurious memories can come into the picture again. Correlated memories naturally involve spurious memories, and spurious memories may be a way to couple the various snapshot attractors together without requiring explicit synaptic couplings to drive the dynamics. We do not know exactly how this may happen, but somehow the spurious memories may act as "filler" to enable the network to move smoothly from one attractor to the next. This scenario may also have repercussions for the way in which conscious or declarative memories are slowly converted into nondeclarative ("attractorless") memories, explaining also how it is that some nondeclarative memories can proceed much faster.

Spurious memories may also be required for thinking. When we think, we mix various memories or attractors together and arrive at new ideas, which once again may correspond to spurious attractors, as they involve subtle combinations of features of stored memories, and here the memories used to initiate the search.

It is well known that our memories are constantly changing, and subject to degradation. We often disagree with people about past events, such as who thought of what first, who did what, or who caught the biggest fish on holiday. Courts of law usually discount evidence given after a considerable time has lapsed. It is not just a matter of forgetting, but memories change with time because of their interaction with each other—that is, through spurious memories. Memories are combined together or interact with each other because they share the same synaptic connections. Spurious memories are the consequence of this sharing. Witnesses may genuinely believe they are telling the truth, even if their testimony is known to be untrue. Spurious memories may then offer an explanation for "false memories." In the simple model considered here, memory number 2 (in figure 4.1) may evolve as a consequence of this intermemory dynamics into the first spurious memory in the third row of figure 4.2, which looks much like it.

SYNOPSIS

Although the Hopfield model is biologically quite unrealistic (as I have argued here and elsewhere in this book), it is remarkable nonetheless, as it has given us fresh insights into how content-addressable memory works in the brain, how creativity may be generated in the brain, and perhaps how the brain is also able to adapt, generalize, learn, and make associations. One wonders if we would have discovered how some of these facets of brain function work if we had started with a more realistic neural network model. Realistic models do not necessarily lead to attractors, and the number of spurious states in them is generally quite low, so we might have not noticed these states and their special properties. Fortune favors the brave.

Animals' brains are made of similar neural tissue and networks as are humans', so if humans have spurious states and creativity, then why not animals too? One would imagine that animals do have spurious states, and use them to learn, adapt, and possibly generalize, but they are certainly not as creative as we are. Humans have developed an ability to utilize their creative spurious states better, possibly aided by their advanced communication skills and the free exchange of information across human society. Some animals (like chimpanzees and gorillas) are known to imitate (which is the first step toward memetic evolution), but this is nowhere as profound as the situation with humans. Another important attribute in the utilization of creativity is the recognition that something useful has been found, which requires a certain knowledge base to appreciate the importance of creative spurious states. This process is aided in humans by their extraordinary mental capabilities and the sharing of public information, and what may be termed a collective human intelligence, or a means to collectively assess new ideas. Humans recognize the importance of these factors and teach each other how to be adaptive, how to think, and how to be creative.

Human creativity greatly facilitates the evolution of memes. It makes sense to imagine that creativity developed first and led to new ideas, such as how to build a fire or how to make (stone) tools, which were conceived as useful and were subsequently copied by others. Something novel or useful needs to be discovered before it is copied. Subsequently, humans developed the ability to imitate each other, through their biological evolution, by selecting those individuals who were the best imitators, as this gave them a survival advantage. This propelled the evolutionary process of memes to what it is today. Memetic evolution clearly benefits from human creativity, but it may also subsequently lead to refinements of the creative process and the accelerated evolution of ideas.

What makes the brain particularly adaptive, associative, and capable of generating new creative states is that it can generate its own memory states, which were not intentionally stored in the system. These spurious states are a natural consequence of the distributed overlapping storage of memory. They are also important for the human ability to learn and to generalize. If the brain were a faithful recording device, like a computer or a tape recorder, there would be no creativity, nor would the brain be able to learn or adapt to new situations. The only thing such a device can (re)produce is the information stored in it. One may ask if we will one day be able to build a device, or computer, that is truly creative and able to think. In my view such a "machine" would need to be imperfect, able to generate its own (memory) states, and capable of utilizing those states. Such a device would most probably need to store memory in an overlapping fashion.

The ability to discover things is dependent on what we already know. Knowledge is refined and extended to generate new ideas. This fits the notion that spurious attractors are the source of these discoveries, as they encompass combinations of memories and knowledge. It is also apparent that the more we look at something, the more we see in it, and the more complicated it gets. As we learn more and more, we realize there is even more to know. The more we know, the more questions we can ask. This endless growth of information and knowledge is a general feature of scientific discovery. An important discovery establishes a new field of research, but as time goes by and more knowledge is acquired and new discoveries are made, this field subdivides into new areas of research. The process continues indefinitely, like a tree of discovery. Our knowledge leads to the discovery of finer and finer details. Take the brain, for example. Today there are a vast number of highly specialized research areas on the brain. In the future, each of these areas will subdivide further. I believe that this phenomenon has an explanation in terms of spurious memories, which according to my thesis are the basis for discovery. The more we know—that is, the more spurious memories that have been converted to stored memories during the process of discovery—the more new spurious memories are generated that may be fruitful for new discoveries.

There is, however, a limit to how much we can discover. Limited by the finite number of brain states, this process cannot continue forever. We can discover only as much as the brain is capable of comprehending. Although a human brain can store only some 10^{13} bits of information, the absolute theoretical limit to how many different states it can generate is more like $2^{10^{10}} = 2^{10,000,000,000}$. In deriving this number I have assumed that one-tenth of all neurons in the brain may be involved with memory storage. Each of these cells can have two states (firing or quiescent), so there are 2 multiplied by itself 10^{10} times, possible states of the brain. This is probably an overestimate, because it would not be feasible to have brain states where only one cell is firing (a group of neurons is required to sustain mutual activity) or all cells are firing simultaneously, for example. A better estimate could be derived by assuming that the only brain states possible are those in which one-tenth of all memory cells are excited. In this case the number of available brain states, with a 10 percent coding level, is the number of different ways that we can choose 10^9 active cells from the total number of 10^{10} available cells. This number still comes out to be something like $2^{4,000,000,000}$.

Such numbers are enormous. Consider the following example. Take a sheet of paper and double it, so there are now two sheets of paper. Now double this again, so there are four sheets of paper. Continue doubling fifty times. If a sheet of paper is one-tenth of a millimeter thick, how high do you think the pile of paper will be after doubling it fifty times? Five centimeters? One meter? One kilometer? The height is 2 multiplied by itself fifty times (or 2^{50}) multiplied by 0.1 mm (paper thickness), which comes out to be a staggering 100 million kilometers. This is almost the distance from the Earth to the Sun, which is so far away that light (traveling at 300 million meters a second) takes 8 minutes to arrive from the Sun. If we were to double this pile a few more times, so we are talking about 2^{54} say, it would go outside our solar system. Now try to imagine how big $2^{10,000,000,000}$ is.

FURTHER READING

The Meme Machine, by Susan Blackmore. Oxford: Oxford University Press, 1999. This book offers an excellent introduction to memes and the evolution of information. Blackmore is the originator of the idea that the human brain is so large because we need to imitate and copy, and that language may have evolved to aid the spread of memes.

CHAPTER 5

The Dreaming Brain

Sleep is universal. Almost every animal has periods of "sleep," when it is generally motionless, vulnerable, and unable to process information from its surroundings. So why have practically all animals evolved this characteristic? It is quite extraordinary that we do not really know the answer to this question, given that sleep is so frequent (occurring every night), so abundant (lasting for most of the night), and so essential (sleep deprivation results in death). It is generally thought that during sleep the brain and body are resting or recuperating, and this may well be one of the most important functions of sleep. Since our metabolic rate is lower during sleep, it may help us live longer. Or perhaps sleep exists to keep us "off the streets" at night, when the situation is potentially more dangerous. Such views are not the whole answer. There is much more to sleep! During some periods of sleep the brain is known to be extremely active, particularly during so-called rapid-eye-movement (REM) sleep, when most dreaming occurs.

This chapter is mainly concerned with the workings of the brain during REM sleep, and the possible functions of REM sleep and dreaming. One theory proposed by Francis Crick and Graeme Mitchison suggests that during REM sleep the brain is engaged in an unlearning or reverse-learning process, which helps it cull spurious and unwanted memories (Crick and Mitchison 1983). Another theory, proposed by the neuroscientist Jonathan Winson, suggests instead that we are relearning (or rehearsing) our memories (not unlearning them) (Winson 1985, 1990). In both of these theories intentionally stored memories are strengthened, and spurious unintentional memories are weakened. Yet another possibility, which is consistent with the suggested elevated status of spurious memories (chapter 4), is that we dream to generate more spurious states (not remove them), to roughen up our "memoryscape" as it were and prepare us for a new day of learning; or perhaps we dream to forget (Christos 1998a).

REM SLEEP

The crudest classification of sleep is in terms of whether or not it is accompanied by rapid eye movements. This leads to the distinction between REM and NREM (non-REM) sleep. NREM sleep can be further classified into four different stages by using an electroencephalogram (EEG) recording of the electrical activity at specific locations on the surface of the brain. These four types of NREM sleep are called stage 1, stage 2, stage 3 and stage 4 sleep.

During the night we sleep in a semi-definite pattern of sleep stages, depending on our age. The sleep pattern for a young adult is shown in figure 5.1. Typically one has four or five alternating NREM and REM cycles throughout the night, with the NREM periods becoming shorter and the REM periods longer as the night proceeds. The first period of REM sleep lasts for about ten minutes and the last (longest) period of REM sleep for about forty minutes. The sleep patterns for infants, children, young adults, adults, and the aged are all slightly different from each other, but they are generally the same within a particular age group. They all have the basic pattern shown in figure 5.1, with alternating cycles of REM and NREM sleep, beginning with NREM sleep. Newborn babies are known, however, to occasionally enter directly into REM sleep at sleep onset.

Stages 2, 3, and 4 sleep are sometimes referred to collectively as slow-wave (SW) sleep because the EEGs for these stages generally have large-amplitude, low-frequency waves (see figure 5.2). SW sleep is usually associated with energy conservation and body restorative functions, since an increase in the amount of such sleep generally follows excessive exercise (Browman 1980; Shapiro et al. 1981). Stage 1 sleep is a transition phase between sleep and wakefulness that usually occurs at sleep onset and sometimes just before waking. Its EEG is similar to that of REM sleep, but it is not accompanied by the eye movements and is not associated with (vivid) dreaming, although it may involve static imagery and some mentation (thinking). Figure 5.1 also shows the relative level of alertness of a sleeper during the night. The deepest stage of sleep is stage 4 sleep, sometimes also called "delta" or "deep" sleep. Contrary to common belief, REM sleep is a relatively shallow form of sleep, and a sleeper is more readily woken from it.

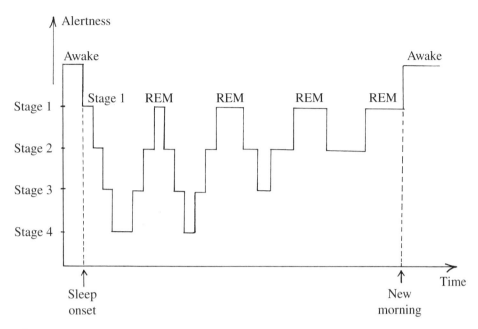

Figure 5.1. The sleep stages for a young adult human during the course of the night. The diagram also shows the level of alertness during each stage of sleep, with the most shallow form of sleep being REM sleep and stage 1 sleep. For more detailed "hypnograms," see Anch et al. (1988) and Kales and Kales (1974).

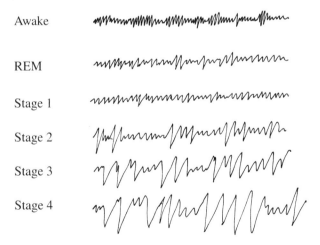

Figure 5.2. Typical (hand-drawn) electroencephalograms obtained from the human brain during the various stages of sleep and while awake.

REM sleep was first identified in 1953 by Eugene Aserinsky (who was at the time a student of Nathaniel Kleitman, a renowned sleep researcher) while observing children during sleep (Aserinsky and Kleitman 1953). Aserinsky had previously noticed that jerky eye movements were characteristic of mental alertness in awake children, so this suggested that during this phase of sleep the brain may be quite active. REM sleep was later observed in adults, and Kleitman would draw the conclusion that it may be associated with dreaming.

During REM sleep, infants are quite active, whereas adult REM sleep is accompanied by muscle-tone relaxation and immobility. It so happens that dream action is blocked in adults by a small group of cells in the brain stem. These cells intercept neural signals to move (generated in the brain) and stop them from being relayed down the spinal cord to the limbs and other parts of the body. When these cells are removed in cats, the cats are seen to physically act out their dreams (see below). These deactivating cells are not properly developed at birth, and this is why REM sleep in infants is sometimes also referred to as "active" sleep. The EEG during REM sleep is relatively flat but jagged, very similar to the EEG in the awake state (see figure 5.2), which is also associated with rapid and jerky eye movements. REM sleep and the awake state are also both characterized by irregular respiration and heart rates.

Another interesting fact about REM sleep is that during it there is actually an increase in cerebral blood flow, or blood flowing into the brain, compared with the other sleep stages and possibly even with the awake state (Meyer et al. 1981). The close similarity of the EEG of REM sleep to that of the awake state, as well as the high cerebral blood flow, suggests that the brain may be particularly active during REM sleep. This is all rather strange, because in adults the brain is practically disconnected from the rest of the body and the environment during REM sleep. It receives very little external input from the senses, and the motor system is deactivated. For this reason, REM sleep is sometimes referred to as "paradoxical" sleep. The brain is isolated from the outside world but seems to be very active. So what is the brain so busy doing during REM sleep? The most logical explanation would seem to be that REM sleep is involved with some sort of internal processing in the brain, possibly involving memory, which is one of its most important functions.

We dream almost exclusively during REM sleep, and almost all REM sleep is associated with dreaming. This fact was ascertained by William Dement (another student of Kleitman) and Kleitman shortly after the discovery of REM sleep (Dement and Kleitman 1957). Dement and Kleitman actually woke their subjects during REM and NREM sleep and asked them if they were dreaming. It is now known that some SW sleep is associated with thoughtlike images, but vivid dreaming with narrative episodes is largely confined to REM sleep. Consequently, REM sleep is often referred to as "dream" sleep. The discovery of REM sleep and its link to dreaming represents the start of a serious scientific post-Freudian investigation of dreaming.

In humans, adults have about one to two hours of REM sleep each day, while infants have about five to eight hours (Roffwarg, Muzio, and Dement 1966). Although infants generally sleep more than adults, the difference between the amounts of REM sleep in infants and adults is still quite significant. Figure 5.3 shows the amount of REM and NREM sleep in humans from birth to old age, using a condensed (logarithmic) scale for age. The brain seems to be endowed with an internal mechanism (probably controlled by hormones) that ensures that the amount of REM sleep needed steadily declines from eight hours each day at birth to just less than one hour in old age. Studies with premature infants have revealed that a fetus has even more REM sleep, and extrapolation of these results suggests that around the thirtieth week of gestation a fetus may engage in REM sleep for almost the entire day (Parmelee et al. 1967). Any serious theory about the function of REM sleep must be able to explain why infants have so much more REM sleep than adults.

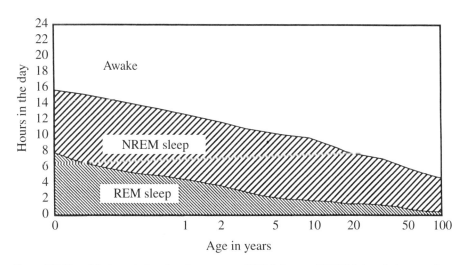

Figure 5.3. Simplified graph showing the amount of REM sleep and NREM sleep in humans from birth to old age, based on the results of Roffwarg, Muzio, and Dement (1966).

THE BIOLOGICAL IMPORTANCE OF REM SLEEP

The biological importance of REM sleep is indicated by the fact that almost all mammals (placental and marsupial) and most birds have it (Allison and van Twyver 1970; Allison and Cicchetti 1976). Other than two species of dolphin, considered to be anomalies (see discussion below), the only other mammal that was thought not to have REM sleep is the echidna, an ancient egg-laying mammal found in Australia, which is low on the evolutionary scale (Allison, van Twyver, and Goff 1972; Mukhametov 1984). As the echidna has a relatively large prefrontal cortex, this would be consistent with theories suggesting that REM sleep is involved with improving the memory storage capacity or the efficiency of the brain (Smith 1902). More recent research, however, suggests that the echidna actually does have REM sleep (Nicol et al. 2000). Other older animals (in an evolutionary sense), such as reptiles and fish, generally do not have REM sleep. Since almost all mammals have REM sleep, it appears that we have evolved with this phase of sleep for some very good reason, possibly connected with survival, the main driving force in biological evolution. For example, if REM sleep improves the efficiency of memory storage and retrieval, this would allow animals to have smaller brains and heads, which could represent an evolutionary advantage. One would hope, however, that the actual reason for REM sleep may be a little more profound than mere head size, since whales and most dolphins have REM sleep, and head size would not appear to be an important issue for mammals that live in water.

It is thought that animals dream during periods of REM sleep, since pets show characteristic facial expressions similar to those observed in human dreaming. In one particular experiment, the part of a cat's brain in the pons/medulla region of the brain stem, which normally deactivates dream actions during REM sleep, was bilaterally destructed (in both the left and right hemispheres). The cat appeared to be acting out its dreams while in REM sleep, stalking and attacking invisible prey in its cage (Morrison 1983; Pompeiano 1979). Based on these results, it would be reasonable to suggest that dream actions are blocked during REM sleep so that we do not hurt ourselves (or our partners) while we dream.

The importance of REM sleep is further supported by the phenomenon of "REM rebound," which was discovered by Dement in 1960. A person deprived of REM sleep one night will generally have twice as much REM sleep on the following night. Continued REM sleep deprivation results in even larger amounts of REM sleep rebound (Dement 1960). It is difficult to continue with REM sleep deprivation experiments with humans (other than with drugs), because after a few days, the subjects refuse to take further part in the experiments, and after a while it is difficult to prevent them from slipping directly into REM sleep when they fall asleep.

In a clever (but cruel) experiment, the physiologist and sleep researcher Michael Jouvet flooded his laboratory and placed cats on small sloping, floating islands (see Hooper and Teresi 1992, 290). The cats had to stand while they slept. However, when they went into REM sleep, the accompanying muscle relaxation caused them to slide into the water and wake up. This prevented the cats from having any REM sleep, but they could still have SW sleep. After a few weeks of REM sleep deprivation, some of the cats died, while the others showed very weird behavior.

Dement also discovered that the time it takes to enter REM sleep (called "REM latency") is reduced after REM sleep deprivation. In other words, a REM-deprived sleeper enters REM sleep earlier than usual. After prolonged periods of REM deprivation, the sleeper may enter directly into REM sleep. Originally Dement found that REM deprivation resulted in certain types of psychotic behavior (hallucinations and delusions), but these findings have not been reproduced, and it is now believed that there are no obvious links between psychosis and short-term REM sleep deprivation or deficiency (Kales et al. 1964).

It is not understood why our sleep is so well organized in alternate cycles of SW sleep and REM sleep. In my view, REM sleep serves some important neurobiological function, as almost all mammals have evolved with REM sleep, REM sleep is an active brain state, and lost REM is recovered during subsequent nights. A possible explanation for the alternating pattern of sleep may be to ensure that we get at least some REM sleep in case our sleep is disrupted. During evolution we may have had to sleep for short intervals to avoid predatory dangers, and this may be why sleep is organized the way it is. It is interesting to note that smaller mammals, which are more likely to be preyed upon, have comparatively short SW/REM cycles. This however does not explain why SW sleep precedes REM sleep or why most of the REM sleep does not occur at the beginning of sleep. Alternatively, the importance of SW sleep may be underestimated. It is known, for example, that stage 4 delta sleep is also associated with a rebound phenomenon and so is probably biologically important.

REM sleep is associated with the mild consolidation of memory (Bloch, Hennevin, and Leconte 1977; Horne 1988; McGrath and Cohen 1978). Experiments with rats have shown that REM sleep deprivation hinders performance of a task the rats were taught a few days before. In these experiments the rats were deprived of REM sleep but were allowed to have the other forms of sleep. It was also found that days on which the rats were required to learn more were followed by nights with increased amounts of REM sleep (Fishbein, Kastaniotis, and Chattman 1974; Smith et al. 1974; Smith 1985).

Similar results have been demonstrated in human experiments (Smith and Lapp 1991; Karni et al. 1994), with more subtle forms of memory. Karni and his coworkers noticed an improvement in a certain visual discrimination task after a normal night's sleep, but this improvement was absent if the sleeper was deprived of REM sleep but allowed SW sleep. Previously acquired visual perception skills were unaffected by REM deprivation, which suggests that REM sleep is important for processing recently acquired information only. (See also the discussion below on dream content experiments using "red goggles.")

Students preparing for examinations generally claim to remember better what they have reviewed on the previous night if they sleep well, as opposed to staying up later, reviewing longer, and sleeping less. It is unclear, however, if this is due to a reduced amount of REM sleep or simply fatigue. There are no experimental results to confirm or disprove this claim. From a personal perspective, if we do not sleep well we tend to feel lethargic the next day; we are unable to think clearly, concentrate, or learn properly. What is particularly interesting is that we feel this way even if we miss only one or two hours of our usual amount of sleep. This may have something to do with the fact that the longest period of REM sleep occurs at the end of our sleep and is what is lost if we wake prematurely.

Generally, experiments with humans do not show profound effects as a result of REM sleep deprivation. These experiments are however quite difficult to perform, and even more difficult to interpret. The fact that REM sleep deprivation seems to have a more profound effect on animals than on humans may be attributed partly to the fact that animals are not aware of the experiments that are being conducted on them and may become distressed, thus exaggerating the results (Horne and McGrath 1984; Horne 1988). Humans who take part in REM sleep deprivation experiments, on the other hand, know that the experiments will continue for only a few days and can refuse to take further part in them.

DREAM DYNAMICS

Even today, people from many different cultures believe that dream content is meaningful and that dreams may contain revelations or messages from God, or make predictions about the future. It is quite difficult to persuade them otherwise, because these notions have been passed on from generation to generation and are often tied to religious beliefs. Dreaming also has a certain mystical appeal to it. Such beliefs are unfounded, however, and scientific evidence suggests that dreams are internally generated and are involved in the processing of the brain's own stored memory.

During REM sleep, a small group of a few thousand neurons located in the brain stem, called the "REM-on" cells, become very excited and persistently stimulate most of the forebrain, which includes the neocortex, the thalamus, and the limbic system (containing the hippocampus and the amygdala) (see figure 5.4). These signals, which seem to be semirandom or erratic, travel via the

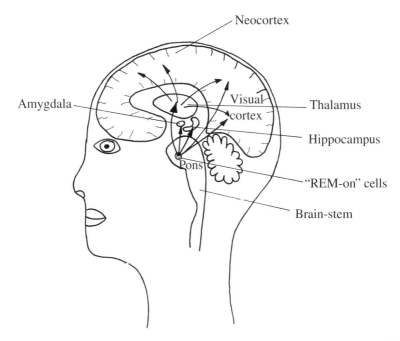

Figure 5.4. During REM sleep, a small group of cells in the brain stem, called the "REM-on" cells, becomes excited and stimulates the neocortex, the thalamus, and the limbic system. According to Hobson and McCarley, dreams are the result of the brain trying to make sense of this haphazard input.

thalamus, through the same channels that are normally reserved for visual and auditory input, from our eyes and ears when we are awake. It seems that the brain has replaced the usual input from the environment with this internal stimulation during REM sleep. This is known as the "activation-synthesis hypothesis" (Hobson and McCarley 1977). The brain is thought to process this internal input during REM sleep in much the same way as it does when we are awake. Hobson and McCarley suggest that dreams are the result of the brain trying to make sense of this noisy input. This could explain why our dreams appear to be so real, why they are somewhat bizarre, and why they represent an intense visual and auditory experience. The fact that the amygdala is also selectively stimulated may explain why dreams are so emotional and are often associated with fear.

During sleep, the brain seems to be influenced by a battle between two groups of competing neurotransmitter systems, the aminergic neurotransmitters (like norepinephrine and serotonin) and the cholinergic neurotransmitters (like acetylcholine) (Hobson, McCarley, and Wyzinski 1975). One or the other of these two chemical systems dominates, with an oscillation period of around ninety to a hundred minutes. REM sleep is associated with heightened cholinergic activity, and NREM sleep with more aminergic and less cholinergic activity. Wakefulness is generally associated with strong aminergic activity. This push-pull mechanism is thought to be largely responsible for the REM/NREM cycles shown in figure 5.1. Hobson and his colleagues suggest that we have REM sleep to allow the brain to replenish its aminergic stocks, which are important for learning. The cholinergic REM-on cells can be turned "on" by the microinjection of cholinergic agonists (drugs that enhance or imitate acetylcholine) and "off" by injecting cholinergic antagonists (drugs that reduce or block cholinergic activity) (Hobson 1990). In this way, REM sleep can actually be induced in an animal.

There are also groups of neurons called the "REM-off" cells, which are aminergically based, that are turned "off" or rested during REM sleep but are usually very active during wakefulness. These groups include the locus coeruleus and the raphe nuclei, which are respectively the main sources of norepinephrine and serotonin, neurotransmitters thought to be important for memory, attention, and learning. While we are awake, the locus coeruleus distributes norepinephrine to much of the neocortex when something is to be learned (Aston-Jones and Bloom 1981a; Hobson 1989). The REM-off neurons are practically inactive during REM sleep (Aston-Jones and Bloom 1981b; Hobson

1989). Although REM sleep shows many characteristics that are similar to the awake state, such as high cerebral blood flow and desynchronized electrical activity, it utilizes a distinctly different neurotransmitter system.

The reciprocal cholinergic-aminergic cycle during sleep may also partially manifest itself during the awake state as a rest-activity cycle (also about ninety to a hundred minutes), which was observed in humans during wakefulness (Kleitman 1963). Daydreaming and periodic tiredness (or diminished alertness) may be associated with periods of cholinergic hyperactivity during the day.

Since the input from the brain stem is semirandom, or quite noisy, this may partly explain why our dreams are so bizarre. This quality is amplified by the fact that some of the brain's inhibitory neurotransmitters (like norepinephrine, serotonin, and GABA) are subdued during REM sleep, putting the brain in a potentially more excitable state. Dreams result from the neocortex (the brain's memory center) trying to make sense of this input, which explains why our dreams are also so personal. The brain-stem input is processed by using the memory stored in the synaptic connections. We dream about things that are relevant to us.

Hobson suggests that many other features of dream sleep can also be explained by the activation-synthesis and reciprocal push-pull theories (Hobson 1988, 1989). For example, the reason why our cognitive abilities are heavily impaired during dream sleep and consciousness is largely absent is because the aminergic system, which is important for attention and learning, is turned off during REM sleep. During dreaming we are frequently deluded, and there is only a mild form of transient consciousness. Some people, in fact, never recall that they dream, which means that they do not experience any form of consciousness during dreaming. Other aspects of dreaming can be accounted for by what is going on in the brain during REM sleep. Take, for example, the fact that during a dream we seem to be unable to escape from our foe; it is difficult to run away, and we inevitably finish up crawling along the ground, clawing ourselves forward with our fingernails. This sensation may arise from the fact that during dream sleep our motor actions are suppressed by the small group of neurons in the brain stem that inhibit the neocortex from sending messages to the legs to make them move.

Dreams seem to involve mainly visual and auditory experiences, with sensations like pain, touch, and smell relatively suppressed. According to the activation-synthesis hypothesis, the visual and auditory channels are the main input channels in the thalamus that are activated by the REM-on cells, while the other channels are not stimulated as much, and sensory input from the body is generally blocked during REM sleep. Hobson also points out that bursts from the pontine reticular formation (or the REM-on cells) seem to excite those parts of the brain that normally process or determine our head and body positions, while suppressing actual input from the head and body. In other words, this set of neurons makes up its own representation of our head and body as we dream. This explains why during dreaming we do not have a true representation of our bodies and where we are.

Since the aminergic neurotransmitter systems are essentially turned off during REM sleep, we do not have full recourse to our memories and logic centers. As noted above, consciousness and self-consciousness are largely absent during dream sleep. This may contribute to the sensation that we are observers. We do not seem to be able to see ourselves properly, during our dreams, and tend to see things happening in front of us, like a movie camera's view of what is going on. We do not actually feel as if we have a body and a 'self' during dreaming. Logic is also largely absent during dreaming. We are almost totally fooled by the bizarre things that we see. There is probably a good reason

Figure 5.5. Dr. J. Allan Hobson, a neuroscientist and psychiatrist from Harvard University, together with his colleagues, has proposed that the neurochemistry of the brain plays an important role in what the brain does during dream sleep. Recently he has also suggested that this same neurochemistry may have implications for consciousness (Hobson 1998). Hobson and McCarley proposed the "activation-synthesis hypothesis" that dreams result from the brain trying to make sense of the noisy input generated in the brain stem. *(Photograph courtesy of Allan Hobson.)*

for this. If we were too aware of what is going on in our dreams, we would wake up more often and would miss out on our important REM sleep. If, during the course of a dream, we become too logical about something or think about what is going on, we tend to either wake up or have a lucid dream (see the last section of this chapter). What may happen in situations like this is that the aminergic neurotransmitter systems gain some control over the cholinergic neurotransmitter system and the brain-stem stimulation.

In chapter 4, I pointed out that drugs may affect or change the "memory landscape" (the memory structure, or the memory store) of the brain. Different attractors or memories may exist for different brain chemistries. This situation can also be induced by natural changes in the brain's neurochemistry, like those that take place during REM sleep. This may explain why we often feel that we have a different identity and different memories during dreaming. We often seem to know the strange characters in our dreams, yet when we wake up they are quite unfamiliar to us. Sometimes we go back into the same dream, either later or on the next night, and feel as if we know everyone in that dream, as if it is a continuation from the previous night. This sensation could have a simple explanation in terms of changes to the brain's neurochemistry, but may however be a simple result of dream delusion; we could just be imagining that these characters are familiar to us.

The set of neurons that control the eye movements during REM sleep (the so-called oculomotor neurons) are located in the brain stem. These neurons are directly activated during REM sleep by the nearby and excited REM-on cells. This explains why dream sleep is associated with rapid eye movements. The activation of the oculomotor neurons is thought not to be accidental, because all mammals have these characteristic eye movements and it would seem odd for nature to develop such a universal but useless feature. There must be some more fundamental reason for the eye movements.

The direction of eye movements is relayed to the thalamus and the neocortex, so they are incorporated into our dreams. This explains the close connection that is observed between dream imagery and the actual direction of our gaze. Our eyes tend to move in synchronization with what we are "seeing" in our dreams (Dement and Kleitman 1957). When we look to the left in a dream, our gaze predominantly shifts to the left. This was discovered by waking people up during a dream and asking them what they were dreaming about and where they were looking, and then correlating their response with the actual eye movements that were recorded previously. It has also been found that people who are blind from birth do not have rapid eye movements, or visual dreams (Berger, Olley, and Oswald 1962). On the other hand, people who have become blind after birth still have rapid eye movements and experience some visual dreams. This seems to imply that the eye movements are an integral part of the reprocessing of visual information, which is not required in those who are blind from birth.

Work with lucid dreaming, or conscious dreaming (to be discussed at the end of this chapter), when a subject actually becomes conscious during a dream, has helped to demonstrate a close link between dream imagery and brain function. If we dream that we are singing, this will activate the same parts of the brain we would normally activate if we were singing while awake. Imagining that we are singing (while we are awake), on the other hand, does not produce any significant brain activity associated with singing.

ON THE FUNCTION OF REM SLEEP AND DREAMING

The function of REM sleep is one of the main unsolved problems in systems neurobiology. It has fascinated scientists for most of the last two centuries, while the meaning of dreams has fascinated humankind for millennia. Here I briefly note some of the main theories of the function of REM sleep and dreaming, and suggest some new interpretations of my own.

Psychological Theories

Perhaps the best-known theories about the function of dreaming ascribe it to psychological need, as first proposed in 1900 by Sigmund Freud (1994) and later revised by Carl Jung (1976). Such theories are commonly accepted by the general community as having some validity, because people seem to dream more often and for longer when they are distressed, but this is actually not the case. We all have roughly the same amount of REM sleep at a given age, irrespective of our state of mind. What may happen instead is that when we are upset we may wake up more often, and this may occur during some of our shallow REM-sleep periods. We may then remember that we were dreaming, gaining the false impression that we dream more often when we are distressed. There is evidence to suggest that we dream about recent memory, so if we are distressed about something we may well dream about it. This is however a natural thing to do and is not necessarily caused by the distress itself.

Freud and others conjectured that there was a close relationship between dreaming and sex. This is also false, and may be explained along similar lines. We may just wake up more often when we are dreaming about sex because it excites us. This may then give us the false impression that dreams are mainly associated with sex.

The notion that dreaming has a psychological basis has been popularized by folklore about the meaning of dreams and is perpetuated by psychoanalysts and dream interpreters. As we have seen, however, dreams are the result of the brain (or the neocortex) trying to make sense of the bombardment of random signals or pulses from the brain stem, so they are generally meaningless. They do however involve our own personal memories and so may appear to have meaning at times. If there were any meaning to dreams, though, it would probably be very private and personal and would generally not be easily amenable to textbook psychological scrutiny or interpretation.

One of the main arguments against the idea that dreaming is associated with psychological need is that infants and children have so much more REM sleep than adults (Roffwarg, Muzio, and Dement 1966). Assuming that REM sleep in infants and adults is not fundamentally different (see however the discussion below), it is difficult to imagine what pressing psychological need an infant may have for so much dreaming. This fact was only discovered in the 1960s and was not known to Freud when he proposed his famous psychological and sexual theories. Freud was also unaware that we all dream for about the same amount of time (at a given age) each day, irrespective of any psychological distress, and that a fetus has even more REM sleep than a newborn. I should point out, however, that there may be different forms of REM sleep and some of the infant REM sleep may be fundamentally different from adult REM sleep. Onc argument in support of this is that the hardware required for dreaming (such as connections in the cerebral cortex, as well as from the thalamus to the neocortex and back) is not developed until one to three months after birth (Hobson 1988, 78).

Memory Consolidation Theories

As noted earlier, numerous experiments (mainly with animals) have shown that learning is (mildly) impaired by REM sleep deprivation, and days of increased learning are followed by nights of increased amounts of REM sleep (Horne 1988). There are a number of theories about the function of dream sleep that suggest we are relearning or reprogramming our memory (Dewan 1968; Geszti and Pazmandi 1989; Winson 1985, 1990). These theories posit that the brain strengthens its hold on recently acquired information or skills during REM sleep. Neurobiological evidence, however, presents some challenges to these direct-learning theories. During REM sleep the aminergic neurotransmitters, which are important to learning, are impaired, and memory consolidation is only mild. A theory that views REM sleep as an unlearning process accounts for mild memory consolidation yet is consistent with the absence of aminergic neural activity.

As a slight variation on these consolidation theories, Winson believes that during REM sleep we (or at least most animals) are rehearsing tasks that are important for our survival (Winson 1985, 1990). The main argument for Winson's theory is based on the so-called theta rhythm. It has been observed that when some animals are performing some important task, which is usually different for different animals but arguably connected with survival skills, they emit an electrical signal from the hippocampal region, called the theta rhythm (Green and Arduini 1954). This signal has a frequency of about four to seven cycles per second. It was discovered that these animals also emit this theta rhythm from the hippocampus during REM sleep, so maybe they are practicing those same skills that are important for survival during REM sleep. The problem is that theta rhythm has not been observed in humans, but Winson argues that perhaps something else replaces it in humans.

Winson's theory leaves the question of why humans dream unanswered, and is hard-pressed to explain many of the other features about dreaming, such as why we often do not remember our dreams. Furthermore, it is known that predators and animals that can sleep safely generally have a higher proportion of REM sleep than animals that are more vulnerable to predation. This seems to contradict Winson's hypothesis, since one would imagine that prey have a greater need to rehearse techniques that are important for survival. One of the main arguments against this theory is that if animals were practicing survival skills, such as how to react to (or fight) a predator, why would they do it during REM sleep, when the body is practically disconnected from the brain and consciousness is absent. The theory is also at odds with the fact that the aminergic neurochemical systems, which as we have seen are intimately linked with learning and attention, are basically turned off during REM sleep.

Winson further suggests that during REM sleep the hippocampus "teaches" the neocortex what it has learned recently. This is an interesting idea. As we saw in chapter 3, memory is slowly converted from relying on the hippocampus to being stored solely in the neocortex. REM sleep could be a good time to relay this information, when both the hippocampus and the neocortex are stimulated by the brain-stem neurons. This idea would explain why our dreams seem to be primarily concerned with recent information (gathered over the last few days), but it cannot explain the bizarre nature of our dreams, or the fact that they are frequently not remembered. Furthermore, there does not seem to be any benefit in learning bizarre dreams. On the contrary, unlearning them would seem to be more beneficial. One would also need to explain how the brain preselects from all memory the important intentionally stored memories it is supposed to relearn or enhance during REM sleep, and why the relearning of only the strongest attractors does not lead to obsession.

Theories suggesting that REM sleep is involved in relearning or rehearsing are based on the loose association of REM deprivation with learning impairment. As noted earlier, reverse-learning or unlearning theories can also account for mild memory consolidation during REM sleep.

Winson's theory is hard-pressed to explain why an infant or fetus would need so much REM sleep. What do infants have to rehearse when they know so little? Winson avoids this dilemma by suggesting that there may be two different forms of REM sleep, a dreaming REM sleep and a REM sleep associated with the development of the brain during infancy. (See the discussion below on the ontogenetic hypothesis.) Infants may slowly convert their structural-building REM sleep into dreaming REM sleep as they develop. It is possible that infant REM sleep may be fundamentally and functionally different from adult REM sleep. The existence of different types of REM sleep in infants may help to explain a puzzling fact about sudden infant death syndrome, concerning the age at death (see chapter 6). I believe that infants do experience some dreaming REM sleep, because they show emotional and facial expressions (such as smiling and sucking) during REM sleep. Children generally purport to have dreams as soon as they are able to communicate, which is generally from about two years of age (Foulkes 1982).

The Crick-Mitchison Reverse-Learning Hypothesis

As noted in previous chapters, a general feature of artificial neural networks is that when patterns or memories are stored distributively in a network in an overlapping fashion, the network generates its own set of spurious memories, which generally correspond to combinations of features of stored memories. According to general folklore, these spurious memories interfere with the retrieval of stored memory and lead to false associations among stored memories. In 1983 Crick and Mitchison proposed that dreaming involves a reverse-learning (or unlearning) process, and that we dream to unlearn the unwanted spurious memories, or "parasitic modes," that the brain generates internally (Crick and Mitchison 1983, 1986).

I will discuss the Crick-Mitchison (CM) theory in much more detail in subsequent sections, but basically it goes along the following lines. During REM sleep, the forebrain processes the semirandom input from the brain stem and arrives at "decisions" or "attractors" in much the same way as it does when we are awake, but now instead of learning the resultant (dream) attractor, the brain unlearns it. The semirandom nature of the input from the brain stem, as well as the fact that a number of the brain's inhibitory neurotransmitters are moderated during REM sleep, automatically puts the brain in an excited state where spurious memories are more readily accessible (or recalled). The unlearning process would then tend to have a much greater effect on spurious memories than on stored memories, and the efficiency of the brain is improved by the general weakening of spurious memories. A consequence of this process is that intentionally stored memories are relatively enhanced, so there is also a mild consolidation of stored memory.

Crick and Mitchison suggest that, in addition to reducing fantasy (which corresponds to spurious memories), reverse learning would also subdue obsession (or the dominance of the brain's mental state by just a few strong memories). Memories with a large basin of attraction are more likely to be recalled during dreaming and will hence be subject to more unlearning or weakening.

There is no direct neurological evidence for the CM theory, other than the fact that the two main neurotransmitters associated with learning (norepinephrine and serotonin) are turned off during REM sleep. To properly test the theory, one would need to examine what is going on at the microscopic level and have a full understanding of what actually happens during learning. Another way to test these ideas is through computer simulations (see below), which offer some support to the notion that unlearning can help improve the storage capacity and retrieval capabilities of neural networks.

The CM hypothesis is very appealing as it offers an explanation of why we do not remember our dreams: we are forgetting them, or weakening their trace in memory. That however is not the only possible explanation. Another, as we have seen, can be given in terms of the fact that during REM sleep the brain is stimulated by the cholinergic neurotransmitter system instead of the usual neurotransmitter systems (norepinephrine and serotonin) associated with memory and learning. We may not recall our dreams because the memory or learning neurotransmitters need to be activated to be conscious of what we are experiencing. We may also have different memory stores for different chemical states of the brain, so another possibility is that the dream attractors, which are attained in the cholinergic chemical brain, may not be attractors in the aminergic chemical brain. It is entirely possible that some sort of learning may also be taking place during the cholinergic-driven REM sleep, but it is not clear how this would affect our awake memories, because these two systems may have different attractors.

Any successful theory of the function of REM sleep should be able to explain why infants have so much more REM sleep than adults. Crick and Mitchison suggest that the extra REM sleep in infants may be needed to reduce the possibility of obsession and fantasy in the developing brain while it makes many new semirandom synaptic connections as it "grows" (Changeux 1986). This notion can be tested by computer simulations on neural network models with varying degrees of synaptic connectivity, to ascertain whether additional unlearning is required to maintain control over obsessional and parasitic modes. Other "adult" theories of the function of REM sleep discussed in this section generally cannot account for the large amount of REM sleep in infants and fetuses. The large amount of REM sleep in fetuses is, however, not well explained in the CM theory, as the most exuberant periods of synaptic growth take place after birth.

Ontogenetic or Developmental Theories

The large amount of REM sleep in infants and fetuses seems to suggest that REM sleep is important in the development of the central nervous system (Roffwarg, Muzio, and Dement 1966). This could involve the decoding of genetic information and the hard-wiring of the brain's neural circuits, which are not fully developed in a newborn. It would seem logical to propose that this genetic decoding takes place during REM sleep, when the brain is periodically stimulated and isolated. Support for this theory comes from the fact that fish and reptiles, which are relatively well hard-wired from birth, do not have REM sleep. The problem is that there does not seem to be sufficient information in DNA (only a few ten thousand genes) to tell the brain how to make its one million billion connections. (See however the discussion and warning about this statement in chapter 2.) Most of the wiring in the brain seems to be done semirandomly, through a growth and pruning process. A variation on the "ogtogenetic" hypothesis is to suggest that REM sleep is involved with the pruning process.

Howard Roffwarg and his coworkers suggested that the only function for REM sleep is in the development of the fetal and infant brain and that it serves no function in an adult, where it seems to have carried over for no apparent reason. The ontogenetic hypothesis is based on the fact that infants and fetuses have much more REM sleep than adults, but it fails to explain why adults have any REM sleep or account for any of the general features of dreaming. Although it may not be obvious from figure 5.3, the percentage of REM sleep to total sleep decreases from 50 percent at birth to about 18.5 percent at age ten years, and then increases steadily up to 22 percent at age thirty, before it starts to decrease again (Roffwarg, Muzio, and Dement 1966). Why would it increase from age ten to thirty? I suggest that this increase may correspond to the increased learning associated with the adolescent and young adult periods. Furthermore, the most rapid neurological development in an infant takes place in the first two years (the so-called critical period), when it has about 3,500 hours of REM sleep. Following this critical period, humans have another forty thousand hours of REM sleep. That is a lot of accidental REM sleep carried over.

At birth, as we have seen, infants have most of the neurons of adulthood, and possibly even more. The only difference is that these neurons are not significantly connected with each other. The majority of new connections in the brain are thought to occur by a semirandom diffusion or growthlike process, and the networks are later refined by a pruning process that eliminates unnecessary links (Changeux 1986). REM sleep may be involved in this pruning process, as the brain circuits can be tested in isolation during it. Culling unwanted connections would clearly help the brain conserve its resources and improve its efficiency. This theory corresponds to a modern day revamping of Roffwarg, Muzio, and Dement's idea. As noted in chapter 2, the human brain still makes many new synaptic connections beyond the critical period, during childhood development and probably beyond, into adulthood. In this case there is a need to have REM sleep extended into adulthood. Note also that the pruning process may proceed in association with a reverse-learning mechanism. The unlearning process may weaken certain synaptic connections to a point where they are no longer required and can be culled.

Replenishment of Neurotransmitters

As we saw earlier, during REM sleep the aminergic centers of the brain, or so called REM-off cells in the brain stem, which distribute norepinephrine and serotonin to much of the brain, are deactivated, and the brain utilizes the cholinergic neurotransmitter system instead. Hobson has suggested that the role of REM sleep may be simply to allow the brain to replenish its aminergic neurotransmitters, which are important for learning and attention. What Hobson presumably means by this is that the aminergic transmitters find their way back to the presynaptic vesicles, preparing the brain for a new day of learning. This is an interesting idea, as we generally feel refreshed, alert, and able to learn after a good sleep. Such a state of mind could also be induced if our memory landscape is roughened during sleep (see below), and it is possible that both of these processes (and other processes like reverse learning) are taking place simultaneously during REM sleep.

Reprocessing Information Collected during the Day

It is known that the content of dreams is predominantly concerned with information acquired in the last few days. This was known to Freud, and has been demonstrated by an ingenious experiment in which subjects were made to wear red-shaded goggles while awake (Freud 1994; Roffwarg et al. 1978; Winson 1985). This red shading gradually infiltrated the dreams of the subjects, and after a few days almost all of the dreams were shaded red. What is particularly interesting is that there was a systematic increase of memories from previous days in the later REM sessions, and these memories progressed from the earliest REM period to the last REM period. When the goggles were removed, the red shading practically disappeared after one or two nights. These experiments suggest that REM sleep is primarily engaged in the processing of recently acquired information.

We may dream to eliminate trivial memory, which is collected during the course of the day and serves no practical long-term purpose. During the day we seem to be able to recall very trivial things that have happened to us, such as how many cups of coffee we have had or how many telephone calls we made and to whom, yet such information from the previous day, or earlier, has been almost totally forgotten or erased from memory. The unlearning of such information could be facilitated by the unlearning process conjectured by Crick and Mitchison. Weak memories would get wiped out just as spurious memories do.

Note that if the neocortex is not excessively loaded or if memories are not stored in a too highly distributed fashion (sparse coding), spurious memories are not such a big issue. In this case REM sleep, through an unlearning process, may be mainly concerned with eliminating unwanted trivial memories that are gathered during the day.

Another idea is that memory from the last few days is temporarily stored in the hippocampus and is relayed to the neocortex only if it is deemed to be important (that is, of sufficient strength). Trivial memories could get lost in this transfer process, which can be conjectured to occur during REM sleep.

As a corollary to the idea that during REM sleep the hippocampus teaches (or relays) its store to the neocortex, one can suggest that REM sleep may be involved with the act of slowly converting declarative memories (which are initially based in the hippocampus and need to enter consciousness) into non-declarative memories (which no longer require the hippocampus and need to enter consciousness). (See also the discussion in chapter 3.)

Graded Memory and Forgetting Theories

Another natural consequence of the reverse-learning hypothesis is that memories are degraded retrospectively, with the most recent memories being the most dominant, since older memories have endured more reverse learning or weakening. Such a situation can clearly be advantageous to animals with REM sleep since recent information is obviously more important to their ongoing survival. The general weakening of all previous memory following sleep may provide a basis whereby newly acquired information obtained during the next day will be given precedence. The enhancement of recent memory may also offer an explanation of why the content of dreams (as well as our daily thoughts) seems to be primarily concerned with recent information (over the last few days).

Continued reverse learning allows a network to effectively forget older unimportant memories and weak irrelevant memories (as well as spurious and parasitic memories). The process of forgetting all of these unwanted memories enhances more recent and relevant memories, which are clearly more important for the survival of animals in a new and changing environment.

Forgetting is unlikely to be the most important function of REM sleep because there are other means by which older memories can be forgotten, such as the decay of neurotransmitters (in which memory is effectively stored) or limitations on synaptic efficacies. (See the discussion in chapters 3 and 4.) In

1983 Crick and Mitchison coined the phrase "We dream in order to forget" (Crick and Mitchison 1983). This was (apparently) proposed only as a slogan in reference to the parasitic modes and was retracted in a later publication, where it was replaced by "We dream to reduce fantasy and obsession" (Crick and Mitchison 1986). Although forgetting may be facilitated by the decay of memory-storing molecules in the synaptic vesicles, it is a natural consequence of the postulated reverse-learning mechanism. I feel that the original statement may not be so inappropriate: we do tend to forget things that we acquire during the day, and sleep is an ideal time and process to do this.

Roughening Up the Memory Landscape

The general negative aspects of spurious memory prompted Crick and Mitchison, as well as other scientists, to look for mechanisms for how neural networks (and the brain) may eliminate spurious attractors. As we saw in chapter 4, however, in the low storage limit, or in more realistic neural network models with low coding levels, spurious memories are actually not so abundant and may not be as harmful as originally thought. I suggested in chapter 4 that spurious memories may be important for the generation of creativity (Christos 1995b). It is interesting to note that in some simulation experiments with palimpsest models (to be discussed later), where the percentage of spurious memories is very low to start with, reverse learning tends to increase the proportion of spurious memories. In this case, REM sleep may well be associated with the generation of creativity in the brain. Such creativity may be important for a mammal to develop new strategies for changing situations. Maybe that is why we often hear people say, "I'll sleep on it," or "I'll get back to you after a good night's sleep." I also pointed out in chapter 4 that spurious memories may be essential for learning new information.

According to this theory, we dream to roughen up the memory landscape so that it is better equipped to learn and adapt the next day. We feel tired and are unable to think after a long day of learning, or reinforcing the same attractors over and over. Our brain is basically obsessed with the same set of attractors. Reverse learning helps to equalize the strongest memories, which in turn may actually increase the proportion of spurious attractors as there are now more strong memories that can be combined together to form spurious memories. This may explain why we are unable to learn or think if we have not slept well.

Creativity and Solution Theories

Dreams have traditionally been seen as a source of creativity, and there are a few historical examples of people having good ideas while they are dreaming. What may happen is that the brain generates a creative spurious state during dream sleep that satisfies the constraints associated with some previous problem of concern to the sleeper, and this may alert and awaken the sleeper.

Otto Loewi claimed he discovered chemical transmission in the brain based on what he dreamed over two nights. On the first night Loewi jotted down some notes after he awoke during a dream, but was unable to make sense of them the next day. On the next night Loewi had a similar dream and rushed into his laboratory, where he experimentally proved that chemical transmission takes place in the brain. Loewi was awarded the Nobel Prize in physiology and medicine for this work. Dmitry Mendeleyev claimed to have discovered the periodic table during a dream in which he saw all of the elements (except for one) fall neatly into place in columns and rows. He awoke to write down the periodic table.

Other great scientists, like Albert Einstein and Niels Bohr, are reputed to have had some of their profound ideas during dream sleep. Apparently Einstein dreamed about his theory of relativity when he was a boy, imagining himself on a sleigh traveling faster and faster, approaching the speed of light. He observed all of the stars fuse with one another, and it was at that moment that he knew there must be a limitation on how fast we can travel. Einstein spent the next twenty years or so working out the mathematical details of his theory with the help of the mathematician Marcel Grossman. Bohr apparently dreamed about his theory of atomic structure, in which electrons orbit the nucleus of an atom in quantized orbits.

The chemist Friedrich Kekulé discovered the benzene ring while (day)-dreaming of six snakes wriggling around until they joined in a ring connecting each other from head to tail. As he imagined the snakes forming a stable hexagonal structure, he realized that six carbon atoms could also form a stable configuration. Some other people who, it has been suggested, dreamed of their inventions are Alexander Graham Bell (inventor of the telephone), Thomas Edison (inventor of the light bulb and many other things), and Elias Howe (inventor of the sewing machine). In his dream, Howe was surrounded by natives who were about to burn him at the stake. The idea of a sewing machine came to him when he observed that one of the natives was carrying an unusual spear that had a small hole in its head. Dreams are a source of artistic ideas as well. Robert Louis Stevenson finished conceiving the plot of *Strange Case of Dr. Jekyll and Mr. Hyde* in a dream.

It is unlikely that generating creative ideas is the most important function of dream sleep, because many more good ideas have eventuated while one is not dreaming. During dreaming our logical faculties are diminished, and it is difficult for us to check the effectiveness of our dream ideas. Sometimes we wake up from a dream with the belief that we have solved a complicated problem only to realize that we were completely deluded and that our "solution" was complete nonsense. We also rapidly forget our dreams as soon as we awake. If creativity is the main function of dream sleep, why do we not remember them, and why are we not conscious during dreaming?

The theory that dreaming may be associated with the generation of spurious attractors, or the roughening up of the memory landscape, suggests that dreaming may be helpful in generating new ideas and creativity for the next day, and that we actually need these states to learn new things and adapt. It stands to reason that if spurious memories are associated with creativity and dreams are predominantly involved with spurious memories, then some of our dreams may contain creative ideas.

From a slightly different perspective, many people claim to be able to think semiconsciously and solve taxing problems while they are dreaming. The last episode of REM sleep seems to be a good time for this, when we are more likely to have a lucid dream. Some thinking also takes place just before we wake up, but this is generally not REM sleep but stage 1 sleep. As pointed out in chapter 4, thinking cannot proceed without spurious memories, and as dreaming seems to be predominantly involved with spurious memories, this may be an ideal time to tap into some of these cognitive states.

Other Less Frequently Entertained Theories

"On Alert" Theory. As noted earlier, REM sleep is actually quite a shallow form of sleep, and a person is more likely to wake up during it. The psychiatrist Frederick Snyder has suggested that regular intervals of REM sleep may allow animals to be on the alert in case of danger. This theory is supported by the fact that smaller animals, which are potential prey, generally have shorter REM/NREM cycles than larger animals, which are potential predators

(Hobson 1989). The lion, for example, which has no predators, has a long REM/NREM cycle, but this theory cannot explain why lions need to have any REM sleep at all, not to mention some of the other features of REM sleep and dreaming.

Acting Out Future Scenarios. Winson and others suggest that in our dreams we are acting out or rehearsing possible future scenarios. In a sense this is true, as we are acting out some randomly initiated input from the brain stem, using our previous experience and memory to arrive at a conclusion.

Some people claim, however, that they can envisage an actual outcome in a dream some days earlier. This is a much stronger claim, and in my view, based on the laws of physics and causality (that time proceeds in one direction only), these claims are, at best, just rare coincidences.

Entertainment Theories (The Night Theater). One theory has it that the purpose of dreaming is one of night entertainment. The vivid, bizarre night movies may be designed simply for our pleasure. This theory is supported by the fact that dreaming is normally a very pleasurable and positive emotional experience and that we often look forward to going to sleep because we will be dreaming. (I do.) The argument against this theory is that although everyone has REM sleep, there are many people who do not dream—that is, they do not recall that they dream. Also, if this is an important function of dreaming, why are we not completely conscious during dreaming, to enjoy every moment, and furthermore, why do infants and children need so much entertainment? It is more probable that dreams are just a "side show" of the brain processing the random input from the brain stem, as suggested by Hobson and McCarley.

Keeping the Brain Warm or Active. One theory suggests that we have REM sleep to keep the brain warm (Wehr 1992). This theory is based on the fact that there is an increase of blood flow into the brain during REM sleep, and a corresponding increase in brain temperature. I believe, however, that this increased blood flow is required to nourish the cells in the brain because they are working so much harder during REM sleep. The brain-warming theory cannot account for any of the other properties of REM sleep and dreaming.

Another possibility following from this idea is that we have REM sleep to periodically stimulate the brain. This would offer some explanation of why our sleep is organized in REM/NREM cycles, but little else.

Uncommon Behavior. Michael Jouvet has suggested that dreams may be necessary for us to practice uncommon behavior or to "act" in ways that are normally unsuitable or undesirable (as mentioned in Hobson 1989). Examples are sexual desire and aggression. This idea ties in with Freud's notion that dreams are associated with subconscious desires and thoughts. One could argue that the brain is stimulated by noisy input during REM sleep so that it can initiate this sort of uncommon behavior. More probably, however, we experience uncommon behavior because of this random stimulation.

Erection Theories. Males generally have erections during or following REM sleep, and "wet dreams" are a general feature of adolescent dreaming. This is another reason Freud thought dreaming was associated with sex, but it is not clear if these erections are actually linked with sexual dreams. Could erections however, have anything to do with the function of REM sleep? For example, does REM sleep prepare males for possible sexual activity during the night or in the morning, when their partner is normally available, or are they due to an "accidental" excitation of those neurons in the brain that are responsible for erections? It would be interesting to know if human females and other animals also experience heightened sexuality during or following REM sleep, and how much sexual activity actually does occur around these periods.

Keeping the Eyes Wet. The ophthalmologist David Maurice suggests that the eye movements in REM sleep are necessary to circulate fluids that surround the eyes (Maurice 1998). This may explain why REM sleep has rapid eye movements and why periods of REM become longer during the night, but it does not explain why the amount of REM sleep steadily declines with age, or any of the other features of REM sleep and dreaming.

Development of Visual Perception. The sleep researcher Ralph Berger suggests that REM sleep may be important for the development of visual perception of depth (Hobson 1989). This theory cannot however explain why REM sleep needs to persist into old age, or any of the other features of dream sleep.

Practicing Motor Skills in Infants. Infants are generally mobile during REM sleep, whereas adults are motionless. This is probably best explained by the fact that the small group of cells that normally stops us from acting out our dreams is not properly developed in infants. Alternatively, one could suggest that these neurons are not developed in infants so that they can practice (or test) their motor connections during REM sleep. This type of infant REM sleep may then be gradually replaced by more conventional adult REM sleep, which is primarily concerned with processing internal memory.

Multifunctional Theories. Many things have evolved in nature that seem to have more than one function or purpose. For example, we have mouths to talk with, to eat with, to drink with, to bite or defend ourselves with, and to kiss with. It is quite likely that REM sleep may involve more than one of the functions I have discussed above, maybe even most of them. It is possible to reconcile many of the theories with each other; some indeed are subtle variations of each other. As we have seen, the reverse-learning hypothesis is consistent with mild memory consolidation, graded memory, and even the pruning hypothesis. Roughening up the memory landscape is also consistent with reverse learning.

No-Function Theory. Some neuroscientists have seriously suggested that REM sleep, in humans in any case, may serve no biological function at all (Daroff and Osorio 1984). This position is taken because REM sleep deprivation over a few days does not seem to produce any significant effects. Crick and Mitchison do point out, however, that there is no obvious effect from food deprivation over a few days, yet we would not conclude that eating is unimportant. Nonetheless there are a number of patients with brain disorders, such as spinocerebellar degeneration, who do not seem to get much or any REM sleep and yet do not show any obvious psychological impairments (Osorio and Daroff 1980). Another patient, with a shrapnel wound to the brain stem, has practically no REM sleep but seems to be conducting a normal life without any obvious psychological or cognitive abnormalities (Lavie et al. 1984). These examples are the Archilles' heel of any theory that assigns an importance to REM sleep and contradict the biological evidence that dreaming serves an important function. In a normal lifetime we dream for about five to six years in total. It would seem illogical to think that REM sleep serves no function at all. One possibility is that the brain of these REM-less subjects may have reverted to some alternative (backup) mechanism that essentially replaces the function of REM sleep.

It is also possible that we may have evolved to a stage where REM sleep is no longer so important. As noted, previously, two species of dolphins, which are highly evolved mammals, do not have any REM sleep. Some of the previous functions of REM sleep (whatever they were) may have been replaced by other brain mechanisms in the course of evolution.

REVERSE-LEARNING THEORIES REVISITED

The reader will have undoubtedly noticed by now that I favor a version of the Crick-Mitchson hypothesis (combined with Hobson and McCarley's activation-synthesis hypothesis and chemical oscillation theory) as the best explanation for what the brain is doing during REM sleep, although I suggest some slightly different interpretations of the usual theme. In particular I suggest that we may dream to forget, to grade our memories, and to actually generate more spurious memories instead of eliminating them.

To briefly recap, during REM sleep, our brain becomes isolated from the environment and our body, and the usual sensory input (such as that gathered by our eyes, ears, and other sensory organs) is replaced by semirandom or "noisy" input from the brain stem. A small group of a few thousand or so neurons, referred to as the REM-on cells, becomes excited and stimulates most of the forebrain, using the same channels in the thalamus through which normal sensory input would pass. From all accounts (EEG signals, cerebral blood flow, and the flow of information or neural charge), the brain seems to be particularly active during REM sleep and appears to be processing the noisy input from the brain stem as if it were real sensory input. This may offer an explanation of the intense realism that we experience in our dreams. We are using the same apparatus to visualize our dreams as we do when we visualize reality. The key difference between the awake state and REM sleep is that the chemical messengers (norepinephrine and serotonin) normally associated with attention and learning are turned off during REM sleep.

Crick and Mitchison have proposed that during REM sleep the brain goes into a reverse-learning or unlearning mode, where processed (brain-stem) input is now unlearned instead of learned, as is normally the case when we are awake. More specifically, the brain arrives at so-called dream attractors, which are now (partially) erased from memory by a process that is the reverse of the usual Hebbian process. (See chapter 3 for a discussion of Hebbian learning.)

Figure 5.6. Dr. Francis Crick, who in 1962 won the Nobel Prize in medicine with James Watson for their discoveries concerning the molecular structure of DNA (deoxyribonucleic acid) and its significance for information transfer in living material, has spent the last twenty-five years studying the brain. In 1983 Crick and Graeme Mitchison proposed that "dream sleep" may be involved with the unlearning of spurious memories. More recently, Crick has been responsible for the resurgence of interest within the neuroscience community on the subject of consciousness. *(Photograph courtesy of the Salk Institute.)*

That is, if a presynaptic cell fires persistently while the postsynaptic cell is firing, that connection will be weakened instead of strengthened.

The CM hypothesis is founded on the notion that, since memories are stored distributively in the brain in an overlapping fashion, the brain generates its own spurious memories. As we saw in chapter 4, in simple mathematical models these spurious memories are generally made of combinations of features of stored memories. There is every reason to expect that spurious memories exist in the brain, as they are a natural consequence of the distributed storage of memory. Distributed storage is important because it allows the brain to work in parallel (unlike conventional computers, which are serial and handle only one instruction at a time) and endows it with the capacity to accept and manage small errors and noise.

Spurious memories are generally considered to be a nuisance since they interfere with the retrieval of stored memory, although, as I have argued extensively in chapter 4, they may be useful (and perhaps even essential) for creativity, generalization, and for learning new information. In terms of the schematic memory landscape shown in figure 3.4, spurious memories correspond to the small valleys on the sides of the larger, broader valleys, which correspond to stored memories. Normally an input is processed by falling into this landscape and then flowing down an incline until it reaches the bottom of the nearest valley, small or large. If the number of spurious memories is large, they can clearly interfere with the retrieval of stored memories. Crick and Mitchison suggest that the main function of dream sleep is to eliminate or reduce the proportion of these spurious memories, which they call "parasitic modes." The idea is that, due to the random stimulation from the brain stem, the brain is automatically placed in a mode in which the spurious memories are more readily accessible. In terms of figure 3.4, random input corresponds to an input from anywhere above. (When we are awake, we generally receive input that is highly correlated with our memory store. We tend to see the same people and do almost the same things every day.) In this case input is adjacent to the minima corresponding to the large valleys. Certain inhibitory neurotransmitter systems are also restrained during REM sleep, and this helps the brain to retrieve spurious memories.

In the CM theory, dream attractors are only partially unlearned, so that intentionally stored memories are not accidentally erased. The unlearning process tends to smooth over the spurious valleys shown in figure 5.7a, to give a memory landscape that looks something like that shown in figure 5.7b. Continued reverse learning may improve the situation further (figure 5.7c), but eventually with too much reverse learning, all memory is lost, and the memory landscape becomes flat, as shown in figure 5.7d (but probably with lots of tiny spurious attractors). These representations are based on numerical results obtained from computer simulation experiments.

Computer simulations with attractor neural network models support the notion that unlearning reduces the proportion of spurious memory and improves the retrieval of stored memory (Christos 1996; Hopfield, Feinstein, and Palmer 1983; Kleinfeld and Pendergraft 1987; van Hemmen et al. 1990). In these simulations a neural network is taught a number of patterns and is then allowed to engage in various sessions of reverse learning. The memory capacity of the network is tested before and after each "dream" session. The testing proceeds by presenting the network with random input, which is processed according to the usual dynamics described in chapter 3, until it arrives at an attractor or fixed state. The experimenter then checks to see if this state corresponds to any of the stored memories or comes sufficiently close to one of them

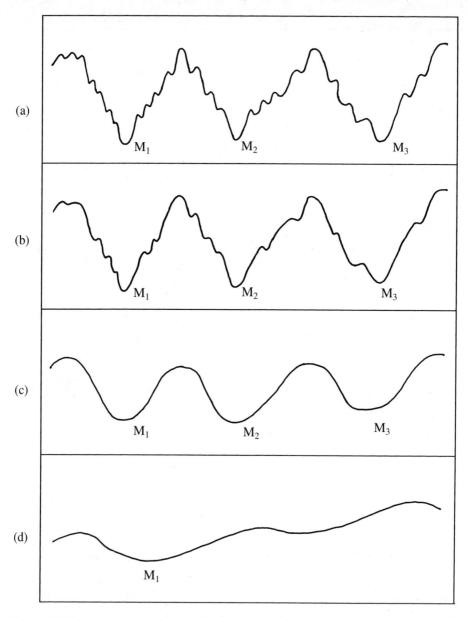

Figure 5.7. Memory landscape representation for a neural network model (without a synaptic bound) undergoing reverse learning: (a) shows the memory of the network prior to reverse learning, (b) shows the memory after some reverse learning, and (c) after additional reverse learning. The effect of reverse learning is to reduce the number of spurious memories (or small valleys) and to equalize the retrieval of stored memories (or the size of the larger valleys). Continued reverse learning eventually wipes out all memory, corresponding to the flattened memory landscape shown in (d).

(say within 5 percent error). If it does, this is counted as a match for that memory. If the attractor does not correspond to any of the stored memories, it is classified as a spurious attractor. The network is then presented with another random input, and the process is repeated. During this testing phase, no unlearning or learning (of the attractors) takes place. The percentage of all random inputs that converge to a particular memory gives an indication of the relative strength of that memory, or the size of its basin of attraction. Simulations show that after a few dream sessions, there is a vast improvement in the retrieval qualities of the network, but as noted above, too much dreaming wipes out all memory.

Simulations have also been conducted with dynamical models, where the network is engaged in alternate sessions of learning and reverse learning as in a real biological system, which learns while awake and presumably unlearns while asleep. It was initially thought that if a network (without a synaptic bound) was to learn and unlearn in alternate cycles, it could function indefinitely without overloading, but this may not be true (Christos 1996; Robins and McCallum 1999; van Hemmen 1997). If the system runs for a sufficiently long time, the synaptic efficacies eventually become large enough to cause the network to overload (Christos 1998b). Some simulations with dynamical models that have a synaptic bound in place (palimpsest models) can trivially avoid the overloading catastrophe, but there is no clear indication of a benefit for intermittent reverse learning (Christos 1998b). This however may be a consequence of our very primitive models, or of the partial nature of the unlearning strength. There is also some uncertainty as to how one should test these networks for what they "know." Should one use random patterns or patterns that are similar to the stored memories?

One interesting result, contrary to naïve expectations, is that reverse learning does not necessarily reduce the proportion of spurious memories in palimpsest models (Christos 1998b). The reason for this may be quite simple. In palimpsest models (or in more realistic neural network models), memories are not stored with equal intensity. Reverse learning tends to equalize the retrieval characteristics of stronger memories (see below) and thus increases the proportion of (stronger) spurious memories that are made of combinations of these stored memories. This is interesting because in chapter 4 I suggested that spurious memories may be the basis of creativity and may be required for learning. If a network had perfect retrieval of stored memories only, how would it arrive at a new attractor to learn something new?

In these models, dream attractors are unlearned with only a fraction (around one hundredth) of the strength of the learning intensity, so that the process does not accidentally erase intentionally stored memory. There are theoretical problems with this situation, as it subdivides the usual synaptic scale. Most of the problems with spurious memories and the overloading catastrophe are caused by the storage of too many memories, or too much information, in the synaptic connections. The finer subdivision of the usual quantized synaptic efficacies can lead to problems with increased interference between stored patterns. It is also biologically unrealistic. Neurotransmitter is released in certain well-defined quanta, and the number of different levels of efficacy is limited. The other problem with partial unlearning is that we would have to "dream" for much longer than we learn. One way to theoretically get around the first problem is to allow the network to unlearn with full intensity but only allow a fraction of the synaptic connections associated with the derived dream attractor to be adjusted. This process works in the case of static models, where the usual weak unlearning process has been tested.

If REM sleep is involved with reverse learning, there are fundamental questions to resolve as to why our sleep is organized into alternate cycles of SW and REM sleep. Maybe something important is happening during SW sleep. As mentioned earlier, episodic (or personal) memory that is initially stored in (or with aid of) the hippocampus is gradually transferred from the hippocampus to the neocortex. Winson suggests that such a transfer may take place during REM sleep (Winson 1985, 1990). Another possibility is that this information is transferred during SW sleep and that the neocortex "edits" it into memory during REM sleep using reverse learning (Crick, private communication). Both of these ideas offer an explanation of why our dreams are associated with more recent memory, while the latter may also explain why our sleep is organized into alternate cycles. The problem with suggesting that the brain is involved in memory processing during SW sleep is that it is not particularly active during this phase of sleep, although the neocortex does seem to receive some weak stimulation from the brain stem during this phase of sleep as well (Rechtschaffen et al. 1972). Since we do not remember most of our SW "dreams," another possibility is that REM sleep is associated with the unlearning of temporal memory (corresponding to continuous narrative dreams), whereas SW sleep is associated with the unlearning of static memories.

Note that, just like the neocortex, the hippocampus can also suffer from the buildup of trivial memories, parasitic memory (fantasy), and obsession, and can benefit from reverse learning. It is interesting that the limbic system, which includes the hippocampus and amygdala, is also heavily bombarded by the random signals from the brain stem during REM sleep.

As mentioned earlier, another benefit of reverse learning is that it subdues obsession. The reverse-learning algorithm tends to equalize the retrieval of strong memories (Hopfield, Feinstein, and Palmer 1983). I believe that this is possibly a much more important property of reverse learning than eliminating spurious attractors themselves, particularly if spurious memories are as important as I have suggested. If a memory has a large basin of attraction, this state will have a much larger chance of being selected during dreaming and will be subsequently reduced or weakened by the unlearning process.

One wonders if neural networks would necessarily become obsessed in the absence of any such mechanism (like reverse learning) that tends to equalize the size of strong memories. We generally learn those attractors that our brain converges to most often, and this would tend to continue to enhance the most dominant memories. This criticism can be leveled at theories which suggest that dream sleep is involved with relearning and rehearsal. One way to get around this problem, which I mentioned earlier, is that since different chemistries are operative in the awake conscious brain and in the dreaming brain, these two brain states may have substantially different attractors or memories: aminergic memories and cholinergic memories. It is not clear what effect learning our cholinergic dream attractors would have on our aminergic awake attractors.

The CM hypothesis is consistent with those experiments that suggest dream sleep is linked to the mild consolidation of recent memory. The greater unlearning of irrelevant (spurious, weak, and old) memory relatively enhances more recent and more important memory. Experiments with REM sleep deprivation are generally associated with only a mild learning impairment. These results are consistent with the Crick-Mitchison hypothesis, which in itself is not a particularly efficient scheme for memory consolidation. The experimental results may, however, be in conflict with theories that suggest dream sleep is more directly involved with learning and rehearsal. Memory consolidation is generally quite weak and has only recently been observed in humans.

Another interesting feature of the CM hypothesis is that it can proceed without any intelligent supervisor (or brain within the brain) to distinguish between true and spurious memories. Random stimulation from the brain stem and subdued levels of the main inhibitory neurotransmitter systems (serotonin, norepinephrine, and GABA) naturally put the brain in a state in which spurious memories are preferentially selected for unlearning (Jacobs and Trulson 1979; Hobson 1989).

Finally, I would like to mention something indirectly related to reverse learning and REM sleep. Numerous reports suggest that a procedure called "eye-movement desensitization" (EMD) has a dramatic beneficial effect in the treatment of patients with posttraumatic stress disorder and some other disorders that involve obsessional characteristics, such as anxieties, phobias, and pain (Acierno et al. 1994; Shapiro 1989; Marquis 1991). To many observers and scientists these claims seem quite unbelievable, since there is no theoretical basis for EMD. (The actual procedure is explained below.) I do not wish to enter into a discussion of whether EMD works or not, but I feel that it is worthwhile to note that there may be a connection between EMD and reverse learning. During REM sleep, the neocortex and limbic system are periodically stimulated by the random signals from the brain stem. The REM-on cells excite the nearby oculomotor neurons in the brain stem, which leads to the characteristic rapid eye movements. The direction of the eye movements is also closely correlated with activity in the left and right hemispheres of the brain. When we process information in the left visual field, we predominantly use the right side of the brain, and vice versa.

In EMD, a patient is asked to hold an active image of a traumatic memory while the therapist induces lateral saccadic eye movements by asking the patient to follow the therapist's finger moving from side to side. It is claimed that the repetition of this simple procedure has the effect of erasing, or significantly weakening, the traumatic memory. It is conceivable (or at least not impossible) that the induced eye movements, which activate the oculomotor neurons in the brain stem, may in turn excite the cholinergically mediated REM-on cells in the brain stem. This may in turn partially activate reverse learning, or cholinergic-based processes that are postulated to occur during REM sleep (Christos 1999). It may then act to weaken the traumatic memory held in attention by the patient. I note, however, that more recently it has been claimed that the EMD technique also works with tapping sounds in place of induced eye movements. It is now called eye-movement desensitization and reprocessing, or EMDR.

DREAMING TO FORGET

Reverse learning helps eliminate weak memories, such as older memories and trivial memories (acquired during the day) that are no longer required. Older memories are progressively diminished, since they have endured more nights of dreaming (here unlearning). This makes the most recent memories, which are arguably more important for survival, more prominent. The general weakening of all previous memory following sleep also provides a basis whereby information acquired during the next day is given priority. I regard these as positive features of the Crick-Mitchison hypothesis. Incidentally, the relative enhancement of recent memory and its general prominence may also explain why our dreams are mainly concerned with more recent memory and the day's residue. They constitute the strongest memories in our brain.

Clearly we have evolved to dream, along with almost all other mammals, for some very good reason. Crick and Mitchison and others have suggested that this evolution produces much smaller and more efficient brains. If REM sleep improves the storage capacity and efficiency of the brain, this may allow animals that dream to have smaller heads. Although I do not dispute that dreaming or unlearning may allow for relatively smaller brains, this does not seem to provide a clear evolutionary advantage. It is hard to imagine why a slightly smaller head should be so important. It would be difficult to explain why humans have such enormous brains, or why aquatic mammals, like dolphins and whales, need any REM sleep, as their head size is irrelevant, since the weight of their heads is supported by the buoyancy of water.

Admittedly two species of dolphin do not seem to have REM sleep, but there may be another explanation for this. Dolphins are highly evolved mammals, and they may have lost their REM sleep for some good reason. Dinosaur fossils have been found that look very much like dolphins, and it is thought that dolphins are the result of around 100 million years of evolution. The bottlenose dolphin has a highly convoluted cerebral cortex. In chapter 6 I suggest that the mysterious sudden infant death syndrome (SIDS) may occur during REM sleep. This raises the question of whether the equivalent of SIDS in dolphins may have something to do with why they no longer have REM sleep. Alternatively, we may find REM sleep in these dolphins later, as in the case of the echidna.

The progressive degradation of older memory is useful for survival. It would be helpful if an animal is suddenly moved to a new environment. The animal could slowly forget older memories that are no longer required, so it can learn more about its present environment. The process of forgetting should arguably be quite slow in case the animal needs recourse to some of its older memories. Older memories that are not required for an extended period will eventually be eliminated completely. Note that important old memories relating to automated functions (like walking, communicating, and seeing) are hard-wired during the critical period of neurological development and are probably not eroded by the reverse-learning process. Nondeclarative memories are probably spared in this way.

Forgetting also enables a mammal to release memories that may be painful or obsessional. For example, it is a common experience that when a relative or a close friend dies, we tend to dream almost exclusively about that person, with subsequent diminishing intensity over a period of a week or two. This example not only shows how our dreams are influenced by our strongest memories, but suggests that dreaming may play a prominent role in weakening and forgetting obsessional memory. If we went through a horrific experience, we would be in constant shock if our memory did not diminish that experience over time.

The CM hypothesis provides a natural mechanism by which the brain can gradually forget older memories as it performs its other important functions. The oldest memories are the most eroded, since they must endure more dreaming or unlearning. My suggestion that forgetting may play a prominent role in the function of REM sleep is not made to the exclusion of the other benefits of reverse learning, and I am not claiming that REM sleep is the only mechanism by which memories are forgotten. At the very least, REM sleep may be particularly useful for the forgetting of trivial daily memories. If forgetting is indeed implicated with REM sleep, this could be tested by seeing if children forget faster than adults do, as they have more REM sleep than adults. The forgetting characteristics attributed to REM sleep may also be tested by appropriate sleep deprivation experiments.

DREAM CONTENT

The CM hypothesis is consistent with common knowledge about dream sleep. It naturally explains why we may not remember our dreams. We seem to be able to recall some aspects of our dreams upon waking, by rehearsing backwards the last few moments of a dream. Otherwise the content of the dream quickly fades from memory. Most psychological theories attempt to explain dream amnesia by invoking the mysterious "subconscious" (or "off-line" processing), which is supposed to be intimately associated with dreaming. This, however, cannot explain why we become conscious during some of our dreams.

"Dream condensation," which was identified by Freud in 1900, is the concept that during dreaming we are frequently confronted with images that seem to resemble a combination of our memories. For example, we may see a person whose face closely resembles a few of our friends and acquaintances. Dream condensation can be explained in terms of spurious memories, which naturally consist of combinations of features of three or more stored memories and are more likely to be recalled or activated during REM sleep.

The bizarre nature of dreams can be explained by the heightened activity of spurious memories during dream sleep. Dreams also have the character that one recalls much older memories (like old friends from school), which are not readily accessible while we are awake. This feature can be explained by the random initiation of dreams, which is more likely to activate older weak memories than when we are awake and receiving recent and more structured and known input.

Some of these features of dreams can also be explained by changes in the brain's chemical state during dreaming, in particular by the subdued activity of some of the main inhibitory neurotransmitters. Serotonin depletion, for example, is linked to hallucinations and bizarre images, and is the same way that drugs like LSD act on the brain (Jacobs and Trulson 1979). Dream amnesia can be neatly explained by the depletion of the neurotransmitter norepinephrine, which is implicated with learning.

Although dreams are initiated by an essentially random input from the brain stem, the dynamics of the brain is controlled by the information stored in its synaptic connections. Even spurious attractors have some connection to stored memories. This suggests that dream content is not completely random and meaningless. Dreams involve memories that are already stored in our brain. There may be personal significance to dream content, but it is generally quite subjective and muddled, and I would not advocate the art of psychoanalysis. It is possible, however, that we may remember (or wake up from) those dreams which are important to our psyche.

One of the main criticisms of the CM hypothesis is that it does not explain the narrative usually associated with dreaming (Hobson 1988; Winson 1990).

When we dream we experience what is happening as a continuous story. There are no stops and starts when the brain reaches an attractor, as it unlearns it before it proceeds to the next attractor. The process is much smoother. This criticism is, however, a little unfair, as the same thing happens when we are awake, and this would be a problem for attractor neural networks as well. Our conscious awareness has a certain flow to it called the "stream of consciousness." The narrative of dreams may result as the brain tries to make sense of the random brain-stem input, in much the same way as it tries to make sense of sensory input while we are awake.

Recurrent dreams are a bit of a problem in the unlearning theory. If a dream is unlearned, why does it recur? Recurrent dreams may be accounted for by the fact that during such a dream we may have been awoken because something in it startled or frightened us. That is why we remembered that dream. As a consequence, the trace of that dream may have been strengthened, instead of weakened, making it more likely to recur in the future (Crick and Mitchison 1983).

As I have mentioned, researchers have discovered that some people can become conscious during dreaming (LaBerge 1981, 1986). Lucid dreaming (to be discussed shortly) generally occurs during the last stage of REM sleep and is quite infrequent, happening about once a month. Consequently, contrary to some suggestions, this phenomenon does not contradict the CM hypothesis that dreams are best forgotten, or that they are mostly forgotten. There is no likely harm if we occasionally remember a few of our dreams. It is just like missing out on a little sleep. Other theories of dream sleep that attempt to explain the reason why we do not generally remember our dreams by invoking the "unconscious" mind are hard-pressed to explain how consciousness can be attained in dream sleep. Even if we are conscious during one of our dreams, we still rapidly forget it upon waking if we do not rehearse its contents immediately. This would seem to suggest that dream amnesia cannot be explained by the lack of consciousness.

POSSIBLE PROBLEMS WITH LOSS OF REM SLEEP

The bizarre and delusional nature of dreams suggests a possible connection between dreaming and psychosis. As noted above, Freud thought that we dream when we have a psychological need, and he once said, "Anyone who when he is awake behaved in the sort of way that he does in dreams would be considered as insane." Crick and Mitchison naturally suggested that schizophrenia may be associated with a lack of REM sleep or the (partial) failure of the reverse-learning mechanism, since this may lead to obsession, and an excess of parasitic modes may result in fantasy and false associations.

Dement originally reported psychosis (hallucinations and delusions) in patients following REM deprivation over a few days, but these results have not been reproduced by other studies (Dement 1960; Kales et al. 1964; Vogel 1968). Schizophrenic patients are also found to respond in the same way as controls (nonschizophrenic patients) to REM sleep deprivation (Vogel and Traub 1968a). Many of these studies, however, were mainly conducted on depressed patients whose condition may be caused by an excess REM sleep pressure in the first place (see below). As mentioned earlier, there are also some patients with brain disorders who have significantly less (or no) REM sleep and do not show any obvious cognitive abnormalities. Other experiments, particularly with drugs that induce sleep, without REM sleep, do not show drastic psychological or cognitive impairment (Wyatt et al. 1971). It is not clear, however, what other effects these drugs have on the brain.

It is reasonable to suggest that evolution may not have allowed short-term REM sleep deprivation to have serious consequences, since we are likely to experience conditions under which we lose sleep. This may explain the somewhat negative results associated with short-term REM sleep deprivation experiments. Consequently, psychological and cognitive damage may only become significant with REM sleep deficit or malfunction over prolonged periods. As mentioned earlier, it is difficult to continue with REM deprivation experiments on humans for more than a few days. They either refuse to take any further part in the experiment or they go directly into REM sleep and it is difficult to deprive them of REM sleep. This in itself suggests that REM sleep may be important.

Based on results that suggest REM sleep may be associated with the mild consolidation and processing of recent memory, it is reasonable to suggest that a REM sleep deficit may initially lead to a gradual decline in recent memory (over the last few days). This memory deficit could then infiltrate into longer-term memory. Such a pattern of memory loss is similar to that observed in Alzheimer's disease (AD)(Christos 1993). It is interesting to note that an as yet unconfirmed study has found that demented patients (mainly of the Alzheimer's type) have about one half as much REM sleep as controls (Allen et al. 1987). AD patients also have large losses of cholinergic neurons in the hippocampus and other memory centers of the brain (Coyle, Price, and DeLong 1983; Davies 1983; Davies and Maloney 1976). The cholinergic system is the brain's chemical system that is operative during REM sleep, and is also involved with short-term memory (Hobson 1990). The question is, could REM sleep loss contribute to memory loss in AD? Another interesting fact is that the amount of REM sleep in humans decreases from about eight hours a day in a newborn to just less than one hour in old age. Since old people have much less REM sleep, they are much more vulnerable to a REM sleep deficit or malfunction. This could explain why AD predominantly affects old people.

Most experiments concerned with determining the function of REM sleep are centered around depriving subjects of REM sleep and noting what effect this has on them. Another approach that can be followed is to instead increase the amount of REM sleep in human subjects or in animals, by the use of drugs, and to observe what effect this has on them (Christos 1993). One interesting drug (actually a natural hormone) is dehydroepiandrosterone (DHEA), which happens to also be depleted in Alzheimer's disease patients (Sunderland et al. 1989). DHEA increases the amount of REM sleep and seems to improve memory (Friess et al. 1995). Some other acetylcholine-enhancing drugs currently being tested (like donepezil, rivastigmine, galantimine, and tacrine) are also thought to improve memory and cognitive ability in Alzheimer's disease patients. This is interesting because the injection of cholinergic agonists (drugs that aid acetylcholine) into the specific pons region of the brain stem where the REM-on cells are located enhances or induces REM sleep (Baghdoyan et al. 1984).

Depressive illness is usually associated with increased REM pressure and is often treated with antidepressant drugs that happen to inhibit REM sleep, although the precise connection between REM sleep and depression is not clearly understood (Vogel and Traub 1968b). This is consistent with the CM hypothesis, since some dementia is naturally associated with depression and excessive amounts of unlearning tend to flatten the memory landscape, as explained earlier, leading to loss of memory.

My main suggestion here is that dream sleep may be involved with a roughening of the memory landscape, and the generation of more spurious memories, so that we are more able to learn (and remember), adapt, generalize, and be creative the next day. If this theory is right, then REM sleep loss or deficit would impede these abilities.

DREAMING TO GENERATE MORE SPURIOUS ATTRACTORS

As pointed out in chapter 4, spurious memories may be more useful than was first thought, and a possible function of REM sleep may be to generate more spurious memories. As noted above, this can take place even in a reverse-learning scenario. As the strongest stored memories are equalized by reverse learning, this may cause the network to generate a broader base of "strong" memories, which could in turn generate more "strong" spurious memories associated with combinations of their features. In this way reverse learning could roughen up our memory landscape and generate more spurious memories, so that we can store new memories the next day, think, learn, adapt, and be creative. This could take place together with the other benefits of reverse learning, such as the weakening of obsessional states, the gradual forgetting and degradation of old memories, the elimination of daily trivial memories, and the weakening of strong fantasies.

In the low storage and low coding limits, which are thought to be applicable to the brain, spurious memories are not particularly abundant, and the suggestion here is that, by dreaming, we are generating more spurious memories that may be required for new learning. Figure 5.8 illustrates how the equalization of strong memories (possibly through reverse learning) could lead to a memory store that has more spurious memories.

The notion that we dream to roughen and generate memory space fits in well with the fact that we need sleep. When we have not had enough sleep, we are unable to think clearly or learn. This may mean that we are unable to form (or entrench) new (spurious/creative) memories that are required for the thinking (or learning) processes. It is as if we were stuck in some large memory states and were unable to venture out of them. When we are lacking sleep, we feel tired and are unable to process new information properly because our memory store has not been roughened enough, and we get stuck in the same old attractors. We may be unable to work productively because we run out of accessible spurious memories.

If spurious memories are themselves intimately linked with associations among the memories they are composed of, one could also argue that generating such associations during sleep may be the way in which the brain actually sorts, classifies, and integrates memories together into longer-term storage.

A roughening of the memory space does not necessarily require reverse

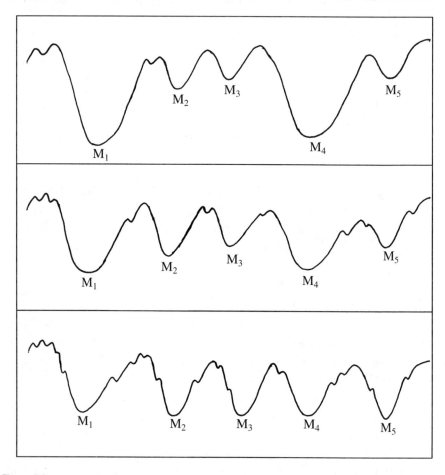

Figure 5.8. A representation of how unlearning, in a neural network with stored memories of unequal intensity, can generate more spurious memories as the retrieval of the strongest memories is equalized. Prior to sleep (a), the memory landscape is dominated by two memories, and there are only a few spurious attractors. As the network experiences reverse learning and equalizes the retrieval of the stored memories more spurious memories develop, as there are now more strong memories that can be combined together ((b) and (c)).

learning. I was keen to combine the two ideas so that we could carry over all of the positive features of reverse learning, such as why we do not generally remember our dreams, and how we may be able to subdue obsession. As pointed out earlier, however, many of the most important features of dream sleep can be explained in terms of the chemical changes that take place in the brain during dreaming, as noted by Hobson and his colleagues. An unconscious form of learning in a cholinergic dreaming brain, which may have different attractors from those of the aminergic awake brain, could also roughen the memory space in some way.

LUCID DREAMING

Lucid dreaming is important to my hypothesis about sudden infant death syndrome (chapter 6) because it highlights the close connection between the mind and the body, particularly during REM sleep, even though the brain and the body are largely disconnected during this phase of sleep.

A lucid dream is basically a dream in which the sleeper somehow realizes that he or she is dreaming and effectively "wakes up," or becomes conscious within the dream (Blackmore 1990; Covello 1984; LaBerge 1981, 1986). In most cases the sleeper is also able to change the course of the dream from that point onwards. Many people have reportedly experienced such dreams. (A survey of my own students reveals that as many as one in three allege to have lucid dreams quite regularly.) Lucid dreaming usually occurs in the early hours of the morning, during the last of the four periods of REM sleep, and usually lasts for a few minutes. The sleeper generally wakes up after a lucid dream. Lucid dreaming, or conscious dreaming, at first sight appears to be a contradiction in terms, especially to those who have never experienced this sensation. In Freudian mythology, dreaming is supposed to be an unconscious process, so how can we become conscious during our dreams?

There are a number of roads to lucidity in dreams. One of the most common is when the sleeper actually wakes up early in the morning, then goes back to sleep and reenters the same dream as before, only this time the sleeper realizes that it is a dream and becomes lucid. One of the key features about becoming lucid during dreaming is the realization by the sleeper that something in the dream is wrong or physically impossible (for example, the dreamer may be flying). Heightened anxiety during a dream can also alert the sleeper that there is something wrong. The sleeper then generally carries out a series of mental experiments to determine if it is a dream. Experienced lucid dreamers, such as those who took part in the experiments mentioned below, generally use this technique to become lucid. One of these researchers was dreaming that she was floating above the ground, when she suddenly remembered that she couldn't float, that floating defies the laws of physics. This made her suspicious that she might be dreaming, so she tried to fly and was successful, and from that point on she became conscious that she was indeed dreaming, and started to control the course of the rest of the dream. Instead of flying over the trees and houses, dodging power lines, she figured that, since it was a dream, she might as well fly straight through all of the trees and poles and through the houses as well.

This particular technique of questioning the feasibility of a dream is difficult to master without considerable practice because during dream sleep the mind is usually disoriented and deluded. The usual mechanisms by which the brain crosschecks the feasibility or reality of what is going on are normally turned off during dreaming. We do not have full recourse to our memories and logic centers. A common experience is to wake up from a dream that seemed to make very good sense (maybe even containing a novel or interesting idea), only to realize almost instantly that it was complete nonsense. It seems absurd that we did not realize we were dreaming. The reason why our logic centers are turned off during dream sleep may very well be so that we do not wake up every time something bizarre happens in a dream. As noted earlier, the bizarreness of dreaming may be an integral part of the function of dream sleep itself.

There are levels of lucidity. In general, when we ask people if they dreamed last night, we are really asking them if they remember that they were dreaming and if they can recall something about their dreams. This involves a certain level of lucidity. Dreaming is normally associated with a transient form of consciousness. We often experience the sensation that we are dreaming without gaining effective consciousness. Experienced lucid dreamers, on the other hand, generally become completely conscious that they are dreaming, and they can even control the content and intent of the rest of the dream. They are also able to send signals (see below) to their colleagues in the sleep laboratory, they can remember instructions given to them before they fell asleep, and they remember details about their dreams when they wake up. At the other extreme, there are some people who claim that they do not dream. Everybody has REM sleep, so what this really means is that they have no recollection of dreaming, as they do not experience any form of consciousness while they are dreaming. Lucid dreaming seems to involve a heightened level of dream consciousness.

On average, people who experience lucid dreams have a few lucid dreams every month or so, but with special techniques and devices one can increase the propensity to become lucid during dreaming. The renowned lucid dream researcher Steven LaBerge claims to have increased the frequency of his lucid dreams almost fourfold by using the following simple procedure, called the mnemonic induction of lucid dreams (MILD) (LaBerge 1981). After waking spontaneously from a dream, LaBerge would memorize that dream, then engage in some other activity such as reading for ten minutes or so before going back to sleep, but this time he would remind himself that he would be dreaming. Typically he would recall images of the previous dream and simultaneously imagine himself lying in bed falling asleep.

There are some other autosuggestive techniques that can help a sleeper become lucid during dreaming. The first is to repeatedly tell yourself during the day that you will become lucid during dreaming tonight. The German psychologist Paul Tholey proposed that you constantly ask yourself, even during the course of the day, if you are dreaming or not (Blackmore 1990). If eventually you ask yourself this during a dream, you may be able to determine that it is a dream. Another idea is to write a big *C* on your hand and constantly look at it during the day. If you find it is not there, you may be dreaming. With sufficient determination and practice one can become more and more proficient at lucid dreaming. Expert lucid dreamers are reported to have lucid dreams almost at will, and can even choose the subject matter before they go to sleep. For example, some researchers are able to dream about swimming underwater simply by looking at pictures with water just before they fall asleep.

There are numerous external mechanical, electrical, audio, and visual devices that have been used to remind sleepers during REM sleep that they are dreaming. Some examples include giving a sleeper an electric shock to the wrist, spraying the sleeper with water, turning on a tape-recorded message repeating over and over, "You are dreaming," or flashing a light on the sleeper's closed eyes. The last-mentioned "dreamlight" machine, which was developed by LaBerge's team, is reported to have been so successful that some people experienced their first lucid dream ever on their first night in the Stanford Sleep Laboratory.

Lucid dreaming has been known and written about for centuries. Aristotle is reported to have said, "For often, when one is asleep, there is something in consciousness which declares that what then presents itself is but a dream" (LaBerge 1986, 21). Until recently, most researchers thought that lucid dreaming simply corresponded to brief intrusions or periods of hallucinatory wakefulness during dream sleep. The work of LaBerge and others on lucid dreaming has shown that this is actually not the case, since sleepers are observed to be in a state of REM sleep while they are experiencing a lucid dream. Lucid dreams are therefore real dreams, in the sense that they occur during REM sleep.

Lucid dreamers at the Stanford Sleep Laboratory are referred to by LaBerge as "oneironauts," which is derived from Greek words meaning "explorers of the inner world of dreams." One of the most important features about lucid dreaming is that the sleeper can communicate back to the real world, or in this case to the other researchers in the sleep laboratory who are monitoring their body. The number of communication channels is limited, because during REM sleep the brain is practically disconnected from the rest of the body except for a few active nerves and connected muscles, such as those that maintain the function of important physiological systems. There are a few other nerve connections that remain active during REM sleep, such as our eye movements.

With some practice, eye movements can be controlled at will by a lucid dreamer. The researchers at Stanford, and independently a group at Hull University in England led by Keith Hearne and Alan Worsley, used this fact to enable a sleeper to communicate with people in the laboratory while dreaming (Blackmore 1990). Typically trained sleepers would move their eyes up and down, instead of sideways, a specific number of times or flick a finger to tell their colleagues in the laboratory that they were lucid. They would also signal when they had started or ended some specific (possibly even prearranged) dream action, such as counting or holding their breath. The colleagues would use these signals to start to record some predetermined physiological activity of the sleeper. On awakening the sleeper would make a detailed report of their lucid dream actions and their estimated time to perform those actions. This information would then be forwarded to an independent observer, who would compare the two reports. What they found was that there is a very close link between what the body does (or attempts to do) and the images and actions that are experienced by a lucid dreamer.

Using these ideas, lucid dreaming researchers were able to answer the age-old question, do our dreams really last for as long as they appear to last while we are dreaming? The common suspicion is that dreams last for a very short time. Often we dream about something like fighting a fire, only to wake up to find that the fire-brigade sirens and bells actually correspond to the alarm clock ringing. This leads us to think that a large portion of the dream may have been created just before we woke up. LaBerge cites a famous example reported by the nineteenth-century dream researcher Alfred Maury, who after a long dream about the French Revolution, dreamed that he was about to be beheaded by a guillotine (LaBerge 1986, 81). As the blade of the guillotine fell, Maury awoke in sheer terror to find, to his relief, that the bed's headboard had fallen on his neck. He concluded that the dream must have been created the instant he was awoken by the falling headboard. Dreams however last for as long as they appear to last (see below), and what may have happened is that the headboard coincidentally fell on his head just as his dream ended, and this may have been combined with his dream.

Lucid-dreaming research has found that dreams last for roughly as long as they appear to last while we are dreaming. The Stanford team conducted experiments in which the oneironauts counted or held their breath for a period whose beginning and end they signaled back to the observers in the laboratory. The time estimated by the dreamer for each of these events was found to be almost identical to the corresponding time recorded in the laboratory, and these times also coincided with the times it would take to perform such tasks if one were awake (LaBerge and Dement 1982a). Although this correlation appears to hold for lucid dreams, one cannot be certain that it is also true for nonlucid dreams. There are however some other indications that it is. As mentioned earlier, the direction of our eye movements during REM sleep is closely linked to the direction (and timing) of our gaze in a dream. There are of course some instances in dreams in which time seems to pass quickly; for instance, you may drive from your home to work in an instant. This situation is similar to how time is portrayed as passing in a movie. Specific dream actions, however, take about the same time as they do in reality.

LaBerge and Dement also found that when a sleeper dreams of counting (for example), this actually activates the left hemisphere of the brain in much the same way that it would if the person were really counting (LaBerge and Dement 1982b). In the same way, dream singing activates the part of the right hemisphere of the brain that is normally used for singing. What is particularly interesting about this is that if you were to imagine (in your head, while you are awake) that you were performing one of these tasks (such as singing or counting), it would not produce any significant shifts in brain activity. Dreaming that you are singing or counting, however, produces almost the same brain activity associated with actually performing these tasks. In short, dreaming is more like actually doing something than like imagining doing it (LaBerge 1986, 96).

The discovery made by lucid-dreaming research that is quite important for my SIDS hypothesis is that we try to act out the content of our dreams, to the extent that this is possible, remembering that during REM sleep the brain

Figure 5.9. An illustration showing the similarity in brain activity when one is actually singing and when one is dreaming that one is singing, compared to imaging singing, when there is virtually no significant brain activity.

is practically disconnected from the rest of the body. The body utilizes those nerves and muscles that are still active while we are dreaming. When an oneironaut dreamed he was swimming underwater, he actually held his breath. When a woman dreamed that she was having an orgasm, her respiration rate increased, and there was increased muscle activity and blood flow in the vaginal region. When an oneironaut dreamed he was writing on a blackboard, he actually twitched the muscles in his fingers that he would have used if he was really writing on a blackboard. At the risk of incriminating myself, I can still recall when I dreamed as a child that I was urinating in my sleep and how this inevitably resulted in a wet bed.

The impulse to act during dreams is probably one of the main reasons why the body and brain are practically disconnected during REM sleep. Otherwise, we might hurt ourselves, or our partners. Recall the experiments with cats, mentioned earlier, that had the parts of their brains removed that usually inhibit the actions seen in dreaming. The cats were observed to stalk around the room as if they were chasing their prey, although they were still in REM sleep, presumably dreaming.

FURTHER READING

The Dreaming Brain, by J. Allan Hobson. New York: Basic Books, 1988. This is a popular book written by one of the world's preeminent neuroscientists, who is particularly interested in REM sleep.

Sleep, by J. Allan Hobson. Scientific American Library. New York: W. H. Freeman and Company, 1989. In this book Allan Hobson tackles the broader subject of sleep. This is another one of those excellent *Scientific American* books about the brain.

"The Function of Dream Sleep," by Francis Crick and Graeme Mitchison. *Nature,* vol. 304, pp. 111–114, 1983. This is the original paper by Crick and Mitchison in which they suggest that dream sleep may be involved in an unlearning process to control parasitic modes.

Lucid Dreaming: The Power of Being Awake and Aware in Your Dreams, by Steven LaBerge. New York: Ballantine Books, 1986. Steven LaBerge is one of the pioneering researchers in lucid dreaming. In this book he not only describes what lucid dreaming is and what has been learned from experiments, but gives practical hints on how you can improve your ability to become lucid during your dreams and how you may be able to utilize dreams constructively.

CHAPTER 6

◊ Unraveling the Mystery of Sudden Infant ◊ Death Syndrome (SIDS)

Approximately one in a thousand of all live births results in a death that is classified as sudden infant death syndrome (SIDS), in which a seemingly healthy infant suddenly and unexpectedly dies during sleep, for no apparent medical reason. SIDS is so poorly understood that it is diagnosed only after every other cause of death has been excluded. To put it quite bluntly, the first and only symptom of SIDS is death.

The first international conference on SIDS was held in Seattle, Washington, in 1963. After some forty-five years of extensive medical research and epidemiology, comparatively little is known about the cause of SIDS, which infants are most at risk, or how to prevent SIDS. While certain risk factors have been identified in the last ten years, the reasons behind them are not completely understood. Our ignorance is even more remarkable considering how many infants die under the classification of SIDS. About five to seven thousand infants die each year in the United States alone. This situation should be contrasted with the breathtaking advances in fighting other diseases. SIDS is arguably one of the deepest mysteries in medical science.

The processes of dreaming and memory may offer a clue to the cause of this mysterious and fatal disease. Basically I am suggesting that an infant dreams it is back in the womb and stops breathing because it did not have to while it was in the womb. After presenting the outline of my fetal memory dream (FMD) theory, I will review what is currently known about SIDS and explain how my theory fits in with these facts. Along the way, I will review some of the other proposed explanations for SIDS. SIDS has attracted many explanations, and only the most important are discussed here. At the end of this chapter I will outline some ways to test the FMD theory further and discuss possible preventive measures to reduce the risk of SIDS.

I believe that the FMD theory is currently the only single cause theory which is consistent with (almost) all of the known facts about SIDS, whereas other theories are generally consistent with only some of the known facts. My theory is particularly useful in explaining what actually triggers the chain of events that eventually results in a SIDS death (Hillman 1991). It is my hope that educational and public-health programs will incorporate my ideas in developing additional preventive measures.

THE FETAL MEMORY DREAM (FMD) HYPOTHESIS

As we saw in chapter 5, during rapid-eye-movement (REM) sleep, the phase of sleep which is normally associated with dreaming, the brain is thought to be processing internal information, or stored memory. The forebrain is periodically stimulated by semirandom input from the brain stem, and dreams are the result of the brain trying to make sense of this noisy input. This is why our dreams are so personal, and involve our own set of memories. Adults generally have one to two hours of REM sleep each night, while newborn babies have about eight hours of REM sleep each day, decreasing to about five hours each day at the age of one year (see figure 5.3). What could these infants possibly be dreaming about? They are dreaming about their own memories, and this inevitably includes their memory of being back in the womb.

In dreaming, our bodies try to act out the content of our dreams as much as this is possible. This qualification is necessary since during REM sleep the brain is practically disconnected from the rest of the body except for a few muscles, such as those that control breathing, the heart, and the characteristic eye movements. Using the technique of lucid dreaming, in which a sleeper becomes conscious during the course of a dream that he is dreaming, researchers have demonstrated that we indeed try to act out our dreams (see end of chapter 5). When a researcher dreamt that he was swimming underwater, his colleagues in the laboratory observed that he was actually holding his breath, while he was imagining to do so in his dream. Other evidence supporting this claim comes from experiments in which those parts of the brain (located in the brain stem) responsible for deactivating most of our dream actions were removed from a cat's brain. The cat was seen to act out its dreams, walking and stalking imaginary prey in a cage while still in REM sleep (see chapter 5).

A fetus has no need to breathe on its own because its mother supplies it with oxygen through the blood. My hypothesis is that an infant, in the course of dreaming about its life in the womb, may stop breathing and subsequently die (Christos 1992; Christos and Christos 1993; Christos 1995a). This simple, but controversial, hypothesis is consistent with all of the known facts about SIDS, including the recognized risk factors.

Another explanation of my theory that does not rely on dreaming per se goes as follows. At birth there is an abrupt change in the way an infant gets its supply of oxygen. In the womb, the fetus is supplied oxygen through the mother's blood, but after birth it suddenly needs to breathe for itself. The brain controls this transition and these two distinct breathing mechanisms. I suggest that the fetal breathing pathway may be "accidentally" triggered during REM sleep. The infant could then just stop breathing and die. This would explain why SIDS deaths seem to be so peaceful, and why the infant does not realize that there is an impending danger. These memories that should be forgotten are more likely to be activated during REM sleep, when the brain is active and processing memory internally, when it is largely disconnected from the outside world (or reality), when it is easily deceived, and when it is being stimulated by semirandom input from the brain stem. The semirandom stimulation from the brain stem during REM sleep makes remote memories more accessible.

In my hypothesis there is either a switch to fetal breathing pathways, or somehow dream content directly influences control over the brain-stem respiratory system. The former is a more appealing scenario theoretically, but the latter is also quite plausible, particularly for a young infant.

Researchers have long struggled to explain why the body's automatic protective mechanisms fail in SIDS. Our airways and blood contain chemoreceptors (as well as baroreceptors) that detect either hypoxia (a fall in oxygen) or hypercarbia (an increase in carbon dioxide). If we try to hold our breath, these chemical mechanisms normally activate reflexes that cause us to automatically resume breathing. These mechanisms should theoretically alert the infant if it was to stop breathing during sleep. Any successful SIDS theory based on cessation of breathing must be able to explain why all of these reflex mechanisms fail simultaneously, and the infant is not aroused from sleep.

The chemoreflex and arousal from sleep mechanisms have a neurological basis, with their control located somewhere in the brain stem. I would argue that during a fetal dream, the infant brain essentially sees itself as a fetus, and the infant protective mechanisms are either not operative or are ignored by the "fetal" brain, as they have no relevance to a fetus. These protective mechanisms are designed for a living, breathing baby. My hypothesis offers a plausible explanation for the collective failure of all of these physiological mechanisms. Other theories based on respiratory cessation need to assert that SIDS infants suffer a massive simultaneous dysfunction of all the chemoreceptor and arousal systems. Researchers have not been able to identify any specific physiological abnormality in SIDS infants that could account for this failure. Even if one could explain why all of these neurological alarms failed simultaneously, one still needs to explain why the infant stopped breathing in the first place. The idea of fetal memory dreams offers a plausible explanation for the trigger of this central apnea, or permanent cessation of breathing.

The question that needs to be addressed now is, do infants dream in the usual sense during REM sleep? For my theory to be plausible, infant dreams do not have to be of a hallucinatory nature, like adult dreaming. An infant can "dream" or recollect those memories of its fetal experience through the same senses that were available to it when it was in the womb.

Many researchers believe that in infants the main function of REM sleep is of an ontogenetic nature, that is, concerned with the development of the brain as dictated by genetics, and maybe with the pruning of synaptic connections. The fact that adults have so much REM sleep presents a problem to this explanation. Infants have about 3,500 hours of REM sleep during the critical period (from birth to two years of age) when most of the brain is connected together, followed by about forty thousand hours of REM sleep in adults. Surely REM sleep must have some other function beyond structuring connectivity and development of the brain. It is widely thought that the main function of REM sleep is concerned with the reprocessing of memory, and that dreaming is a general feature of this type of REM sleep. (See chapter 5 for details.)

We know that children do dream, since they tell us as soon as they are able to communicate, around the age of two (Foulkes 1982). Infants also appear to dream because they show facial expressions, such as sucking and smiling, that are normally associated with dreaming. I do not dispute the fact that some form of brain development may take place in infants during REM sleep, because this is an ideal time for the brain to test its circuits, when it is practically disconnected from the body and is receiving virtually no external sensory input. By the same token, this testing of the circuits could also reinitiate fetal breathing pathways.

The most likely scenario, in relation to REM sleep, could be that infants are born with mostly "structural" REM sleep, but as they get older this type is gradually replaced by the more usual dreaming form of REM sleep found in adults (Winson 1985). I will return to this point below when I discuss the mystery associated with the age at death of SIDS infants. This dual role of REM sleep may go some way towards explaining why infants seem to be protected from SIDS during the first month after birth. It is also known that some of the machinery required for dreaming (or the recollection of memory) is not properly developed in a newborn.

There are other developmental aspects of dreaming and REM sleep that may be of importance here. Individuals can normally realize, with experience, that they are dreaming, especially when a threatening (or embarrassing) situation arises. Children take a few years to develop a sense of when, for example, urinating in a dream will inevitably lead to a wet bed if they do not wake up and go to the toilet. Such skills are not expected to be properly developed in a young infant. In the same way, an infant's ability to "realize" the necessity to breathe may be impeded as a result of its immaturity.

During REM sleep infants are quite active, unlike adults, who are generally motionless. This is because in newborn infants the neural circuitry, or inhibitory systems, that disable "dream action" have not been properly developed yet, or maybe the movements during REM sleep are required to develop an infant's motor systems. As noted in chapter 5, REM sleep was actually discovered by noticing that infants became periodically restless during the night, and this is why it is sometimes also referred to as "active" sleep. The lack of motor inhibition in infants during REM sleep may make them more susceptible to acting out their dreams. The neural system that deactivates dreams may also be dysfunctional in some SIDS infants.

In addition to the development of conventional breathing after birth, an infant needs to develop the neural circuits that are responsible for detecting and reacting to a sudden lack of oxygen, or excess of carbon dioxide. It is unlikely that these mechanisms would be fully developed in a newborn since they were not required in utero, nor can they be properly tested or developed until the infant is born. (The degree of chemosensitivity [such as ventilatory responses to elevated CO_2] in rats, for example, is known to improve with age [Wang and Richerson 1999].) As the infant gets older and its neural circuitry and control mechanisms are better developed, potential dangers can be more readily avoided, either by neurochemical reflex mechanisms or by volition. Arguments of immaturity can be used as a partial explanation of why the incidence of SIDS declines with infant age (after peaking at two to three months), but the actual rate of decrease (which is exponential) is much faster than they can account for. My hypothesis does not rely on these arguments, since it suggests that the incidence of SIDS should decline naturally with age because the infant's memory of being in the womb is forgotten with time. The idea that SIDS infants have a pronounced developmental immaturity is hard-pressed to explain the hiatus of SIDS during the first month after birth (see below).

It is known that a fetus actually practices respiratory movements in utero and that these periods can be quite substantial. This does not discredit my theory, however, because these movements are intermittent (actually occurring mainly during fetal REM sleep) and do not exist for the entire period of gestation (Marchal and Droulle 1988). There is plenty of scope for an infant to dream about times in the womb when it was not practicing respiratory movements.

If fetal memory is such an important part of infant memory, then one might ask why many more babies do not die of SIDS, as they are likely to dream about their sensations in the womb. I would suggest that there may be a number of other factors involved that determine whether a fetal dream becomes dangerous. Maybe the fetal breathing pathway needs to be excited in some specific manner, maybe SIDS infants are unable to relegate their fetal memories related to breathing, or maybe they suffer from some sort of neural dysfunction. I am not suggesting that all fetal dreams are fatal and result in SIDS.

There is an interesting twist to the FMD theory. If REM sleep is indeed involved with reverse learning and in particular with some aspects of forgetting, as was suggested in chapter 5, then it is quite ironic that on the one hand, REM sleep may initiate potentially fatal fetal dreams, and on the other hand, it may play an important role in forgetting or erasing fetal memory.

One of the greatest risks associated with SIDS is allowing an infant to sleep in a prone (or face-down) position. Asking parents to sleep their babies in a supine (or face-up) position, or on the side, has almost halved the incidence of SIDS in developed countries. In my view, my theory is the only theory that can properly account for this risk factor (see discussion below).

There are other theories about the cause of SIDS that are similar to mine. At birth, infants have a primitive instinct called the "dive reflex." When a baby is dropped into water or when its face is immersed, it automatically knows that it should hold its breath. Some researchers have suggested that SIDS infants resort to this mechanism during sleep, stop breathing, and subsequently die (French, Morgan, and Guntheroth 1972; Lobban 1991). French and his coworkers found that a cold wet stimulus induced struggle-free apnea in sleeping infant monkeys. Lobban suggests that the dive-reflex mechanism may be triggered if an infant is unable to tell the difference between water and fabric. However, in order to explain the finding that the risk of SIDS is increased if an infant is placed to sleep in the prone sleeping position or in a thermally stressed environment (see next section), this theory needs to be supplemented with the proviso that the infant confuses "warm" fabric with "warm" water, although the dive reflex is generally associated with cold water. The dive-reflex theory does not properly explain some of the other features of SIDS (to be discussed in the next section), such as why it happens during sleep, other than to suggest that this may be an appropriate time to confuse the brain that its head is immersed in water. One could supplement the dive-reflex theory by suggesting it is triggered by an infant dreaming it is underwater, or even back in the womb, where it was surrounded by embryonic fluid, but then one may as well suggest that the reason why the infant stops breathing is because it did not have to breathe when it was in the womb.

Incidentally, French, Morgan, and Guntheroth first speculated that the lack of arousal in SIDS may have something to do with a reversion to a fetal state, in which the absence of respiration was of no consequence (French, Morgan, and Guntheroth 1972). The FMD theory offers an explanation for the cause of the initial sleep apnea, and why fetal memory is instigated.

Numerous theories about SIDS suggest a link with sleep apnea, of which there are two basic types. "Obstructive" sleep apnea occurs when the flow of air into the lungs is obstructed in some way, such as by the collapse of the airway passage. Obstructive sleep apnea in adults is usually associated with snoring. "Central" sleep apnea occurs when the brain "forgets" to breathe, or to send the appropriate signals to the respiratory muscles. Theories based on obstructive sleep apnea, which may result from previous infection or nasal occlusion, are not taken seriously nowadays. Theories based on central sleep apnea suggest that SIDS results from some sort of malfunction of the central nervous system. Primary sleep apnea may also lead to hypoxic sleep apnea, a deeper form of sleep apnea resulting from a deficiency in oxygen.

In the FMD theory, central sleep apnea, followed by permanent respiratory cessation, is the main gateway to SIDS. Brief periods of sleep apnea are quite normal in infants, and apnea is quite common in adults (Southall et al. 1980). Although apnea generally occurs more during REM sleep, Guilleminault and his coworkers have found that the most prolonged periods of sleep apnea (lasting for more than ten seconds) predominantly occur during non-REM sleep (Guilleminault et al. 1975). This study was conducted by examining infants who had suffered a so-called apparent life-threatening event (ALTE), having previously experienced a prolonged apneic episode that was thought to be linked with SIDS, sometimes also referred to as a "near-miss SIDS." These findings would seem to contradict the FMD hypothesis, which suggests that SIDS occurs during REM sleep. Other findings, however, suggest that sleep apnea per se is not predictive of SIDS, and there is no direct proof that ALTEs are aborted SIDS cases (Southall et al. 1982). Furthermore, according to the American Academy of Pediatrics Task Force on Prolonged Infantile Apnea, "Prolonged apnea rarely leads to mortality, the vast majority of infants with prolonged apnea are not victims of SIDS, and most SIDS infants were never seen to have prolonged apnea prior to the terminal event" (Little et al. 1985). The apneas observed by Guilleminault were also mainly of the obstructive type. His data may not support the FMD theory, but it does not rule out the existence of a rare form of prolonged (and possibly fatal) central sleep apnea occurring in REM sleep, as suggested by the FMD theory. Who is to say that prolonged central sleep apneas, greater than a hundred seconds (say), do not occur mainly during REM sleep?

WHAT IS KNOWN ABOUT SIDS?

In this section I will enumerate the known facts about SIDS. Under each of the listed topics I will discuss some of the other theories about SIDS that are specifically related to this fact, and I will show how each of these facts is consistent with the FMD hypothesis. I will also consider how some of the listed facts influence each other.

Since the cause of death in SIDS is not exactly known, there is obviously only a limited amount of pathological information about it. Most researchers believe, however, that the cause of death is related to either respiratory cessation or heart failure (mainly the former), combined with the failure of the associated reflex and arousal mechanisms. Consequently, pathologists have been particularly interested in examining the lungs, heart, and some specific regions in the brain stem, looking for any abnormalities that may be specifically associated with SIDS infants. This information has mostly been obtained at postmortem. Some other medical peculiarities about SIDS infants have been gleaned from recorded medical information that was gathered from a group of infants while in hospital, some of whom subsequently died under the classification of SIDS. This information includes recorded sleep, brain, heart, and breathing activity.

Other information about SIDS comes from the statistical analysis of data involving SIDS infants. This includes previous medical information as well as information about the infants' sleeping environment, family medical history, and family practices. Some of the epidemiological information that has been gathered in this way has shaped theories about SIDS and has helped save many lives. One of the main achievements of these epidemiological studies has been the discovery of certain risk factors associated with a higher incidence of SIDS. The prone sleeping position, for example, increases the risk of SIDS by a factor of three or more. Knowledge of this has helped reduce the incidence of SIDS by one-half and has saved the lives of tens of thousands of babies over the past ten years.

No Known Cause of Death

There is no known cause of death in SIDS, and there are no typical pathological findings associated with it. Later in this section I will detail some peculiar medical findings associated with SIDS infants, but what is usually the case is that not every SIDS infant demonstrates these pathological characteristics, so they are not typical and hence cannot be used as markers to identify SIDS, or infants most at risk.

As a result of this ignorance, SIDS is diagnosed only after every other cause of death is excluded during an extensive autopsy. This includes a rigorous investigation of the death scene and a detailed review of the medical history of the baby and its family. A SIDS death is recorded only after a very exhaustive investigation. Having said this, I should point out that it has recently been discovered that some deaths previously classified as SIDS were actually suffocation murders (see below). Such cases are however very few, at least today, with advances in pathology and the rigor of investigative procedures. There are clearly many infant deaths every year whose cause is genuinely not known. This is quite remarkable given that tens of thousands of babies die each year in industrialized countries as a result of SIDS, and the rigor in which these deaths are investigated.

The FMD theory suggests that SIDS may be so hard to diagnose because it is, in a sense, in the mind of the infant.

From all accounts babies seem to die peacefully in SIDS. There are no signs of a struggle or anything untoward. Some babies have even died while sleeping in their mother's arms. As noted earlier, most researchers believe that the primary cause of death in SIDS is related to respiratory cessation; strangely, though, SIDS infants do not typically display the pathological signs, such as cyanosis (the bluish purple discoloring of the skin and lips caused by a lack of oxygen in the blood), that usually accompany respiratory seizure. In the FMD theory, an infant may stop breathing without raising any (neurological) alarms that would cause these pathological traits. As far as the infant's brain is concerned, there is nothing wrong and there is no need to struggle to breathe again, as in its dreams its mother is supplying it with oxygen. I suggest that the lack of cyanosis in SIDS may be intimately linked to the subsequent failure of the infant's chemoreceptor systems to alarm the infant of a need to breathe, or a need to struggle for air. Suffocation can sometimes also occur without any of the classical signs associated with it, and it is difficult to rule out intentional suffocation at autopsy. Indeed some suffocation murders were originally misdiagnosed as SIDS (Dix 1998; Firstman and Talan 1997). These deaths were presumably not accompanied by the usual symptoms of suffocation, such as cyanosis and a swollen purple face.

High Mortality Rates

SIDS is the second most common cause of infant death in the first year of life, accounting for around 15 to 20 percent of all deaths. The most common cause of infant death is congenital anomalies (existing at birth), and most of these deaths occur in the first week after birth. SIDS accounts for the majority of postneonatal deaths (that is, from six weeks after birth up to one year of age), accounting for 30 to 40 percent of all deaths (see figure 6.1).

In developed Western countries, the incidence of SIDS is about one in a thousand of all live births. Up until about ten years ago, before parents were advised not to sleep their babies in the prone position, the incidence of SIDS was actually almost twice as high as this rate. The number of genuine SIDS deaths is difficult to estimate in developing countries because these countries are generally not equipped with the medical infrastructure to conduct extensive autopsies on infants, and most infant deaths are not fully investigated. A rough estimate would suggest that somewhere around fifty to a hundred thousand babies die from SIDS around the world every year.

Other than in 1992, when sleeping positions were altered, the number of SIDS deaths each year has remained fairly constant, while the number of infant deaths from other causes has shown a steady decline with time, as our knowledge and medical expertise have grown.

A Sudden and Unexpected Death

A SIDS death is normally completely unexpected. Prior to death, the baby is usually considered to be in excellent health. Numerous studies have failed to identify any significant correlation with the baby's health immediately preceding SIDS (Ford et al. 1997; Taylor et al. 1996). This means that we are unable to predict which infants are most at risk of SIDS. There are, however, some preventive measures, such as avoiding the prone sleeping position, that can be taken to reduce the risk. Risk factors associated with SIDS will be discussed in more detail later.

In the early days, scientists suspected that SIDS was caused by some pathogen, a bacterium, virus, or poisonous substance. No such specific agent has been found to this day, although one cannot rule out that frequent respiratory infections can contribute to SIDS. In any case, pathogenic theories are unable to explain some of the other features of SIDS, such as the age-at-death distribution or why the prone position is so much more dangerous than other sleeping positions. One of the main reasons it was thought that SIDS might be caused by an infection is that the incidence is known to be higher during the winter months (see below), when colds and sickness are more prevalent.

The FMD hypothesis does not assert that there is any link between an infant's physical health and SIDS, although there may be some indirect contributing factors. What is really strange about SIDS, however, is that environmental factors, like sleeping position and family social practices, impact heavily on the risk of SIDS.

While They Were Sleeping

SIDS occurs during sleep. There are however a few old reported cases in which infants have supposedly died while they were "seen to be" awake (Adelson and

Kinney 1956). There are also some cases of SIDS that have been reported to have occurred just after a baby finished breast-feeding, and in many cases where the baby was still in its mother's arms. In spite of these incidents, most researchers believe that sleep is universal in SIDS. Consequently they have been particularly interested in sleep, the various stages of sleep, and the reflex and arousal mechanisms during sleep.

The FMD hypothesis asserts that babies not only die during sleep, but during REM sleep. Infants, like adults, usually enter slow-wave sleep before they go into REM sleep, but they sometimes enter REM sleep directly at sleep onset (Anch et al. 1988, 45; Hobson 1989, 73). This may explain the few odd cases in which it has been inferred that an infant was awake when it died, later classified as SIDS. These deaths may have occurred just as the baby was falling asleep.

Most SIDS deaths are reported to occur in the early hours of the morning, and during the longest sleeping period, which tend to coincide (Bergman 1970). This is another one of those puzzling facts that most researchers ignore, or simply argue for one reason or another that infants are more vulnerable during that part of sleep, possibly because they are in deeper sleep. That is actually not the case, since our deepest sleep occurs at the start of sleep, and the shallowest sleep, which is REM sleep, occurs mostly in the early hours of the morning (see figure 5.1). The propensity for REM sleep increases the longer we are asleep. These facts fit in nicely with the FMD hypothesis, which does not exclude the possibility that infants can die during the other REM stages of the night. It is just that there is more REM sleep in the last part of sleep and one consequently expects that most deaths would occur at this time.

No Boundaries

SIDS has no geographical, historical, or cultural boundaries. It has always existed. Throughout most of human history, infants slept with their mothers, and what may be presumed to have been SIDS deaths were thought to have occurred because the mother had overlain the child during sleep and the baby had suffocated. There is even a reference to SIDS in the Bible, in the Book of Kings: "And this woman's child died in the night because she overlaid it" (as noted by Guntheroth 1995). The fact that SIDS has probably always existed suggests that it is not due to a pathogen or an infection and that it is not directly linked to cribs or crib mattresses, which have been used for only the last hundred years or so. This is one of the reasons why many researchers now prefer to use the name "SIDS" instead of "cot death" or "crib death."

SIDS occurs all over the world, although the incidence varies quite dramatically from one country to another. In Japan, Hong Kong, Finland, and Sweden, the incidence of SIDS is quite low (approximately 0.3, 0.3, 0.6, and 0.5 deaths per thousand live births, respectively), while in New Zealand and the United States it is quite high, at around 1.5 deaths per thousand live births, based on 1996 figures. In Christchurch, a city on the South Island of New Zealand, the SIDS rate was reportedly as high as eight in a thousand in 1986 (Valdes-Dapena 1988). Some of these differences can be attributed to environmental factors, like climate, and to cultural practices, such as the way that infants are put to sleep (see discussion below).

The Age at Death

The age-at-death distribution (see figure 6.1) shows that the incidence of SIDS is generally at its highest level during the second and third months after birth, with approximately 50 percent of all SIDS cases occurring during this period (Goldberg et al. 1986; Peterson 1988). Some 80 to 90 percent of all SIDS deaths occur within the first six months after birth. The incidence of SIDS is also known to decrease exponentially with age after its peak, and only 3 percent of all SIDS cases occur after one year of age. After two years of age the risk of SIDS is negligible. The current definition of SIDS advocated by the U.S. National Institute of Child Health and Human Development (NICHD) describes SIDS as the unexpected death of an infant less than one year of age, with an unknown cause (Willinger, James, and Catz 1991). Ignoring the low incidence of SIDS during the first month, the exponential decay in the incidence rate after the peak around two to three months suggests that SIDS may be intimately linked to the normal development of an infant.

The FMD theory can easily explain the exponential decay of the incidence of SIDS as a function of age by the fact that an infant's memory of being back in the womb decays with time. An exponential decay of memory is equivalent in mathematical terms to a constant multiplicative (or compound) decay of memory or the synaptic efficacies. This is a reasonable model of memory decay. If forgetting is implicated with dream sleep, as in the reverse-learning hypothesis, this leads to an exponential decay of memory.

Another factor that goes some way towards explaining the decay in the number of SIDS deaths with age is the maturation of brain-stem neural network systems that control cardiorespiratory functions and chemorespiratory

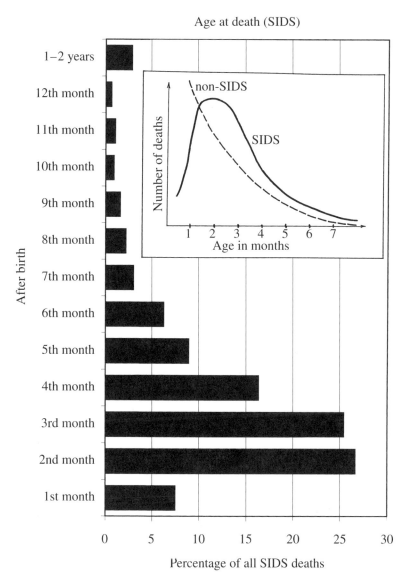

Figure 6.1. The age-at-death distribution for SIDS infants. Data was taken from Goldberg et al. (1986). Note that the incidence of SIDS is relatively low during the first month after birth, is highest during the second and third months, and decreases exponentially after the third month. The inset shows the incidence of SIDS deaths compared to other deaths (non-SIDS) as a function of infant age. Other infant deaths generally peak at birth and decay immediately thereafter. (1988).

reflexes. When an infant is born, as we have seen, it has all the neurons of adulthood, but these neurons are not properly connected. It is conceivable that the neural circuits in an infant's brain, which are concerned with the control of these important neurological functions, are still being developed from birth. It has been observed that SIDS infants (as a group) tend to have a delayed maturation of the brain stem, but this is not general, in that not all SIDS infants have been observed with this abnormality, so it is probably a contributory rather than a causal factor. Theories asserting that SIDS is caused by a brain-stem immaturity are unable to explain why the incidence of SIDS is so low in the first month after birth and do not offer any plausible explanation of the cause of death, or why these infants may have suddenly stopped breathing in the first place.

For some reason, a newborn baby seems to be protected from SIDS in the first month. This puzzling situation is quite different from what happens with other infant deaths, which generally peak shortly or immediately after birth and decline thereafter. Plotting the number of infant deaths that are SIDS and non-SIDS yields a graph that looks something like the inset in figure 6.1 (Peterson 1988). This inset also shows that SIDS becomes the main cause of infant death two months after birth.

The sparing of infants during the first month is a serious problem for most other SIDS theories. One suggestion is that infants are protected during the first month by a very effective "gasp reflex mechanism," which is the mechanism that enables a newborn to gasp for its first breath (Guntherop and Kawabori 1975). The FMD hypothesis offers a possible explanation, based on the neurobiology of dreaming, of why the incidence of SIDS is so low during the first month. There are some indications that the REM sleep experienced by adults and infants is different in some ways, and that infants may have two forms of REM sleep, as discussed earlier. REM sleep in infants, particularly in newborns, is characterized by active movement, while in adults REM sleep is generally motionless. Infants also have a much higher proportion of REM sleep. It is generally thought that most of an infant's REM sleep (particularly for a newborn) is concerned with the development of the brain, and that the usual adult dreaming-type REM sleep develops later. It is conceivable that infants may have two forms of REM sleep, a structural REM sleep that aids brain development and the testing of neural circuits and a dreaming REM sleep as experienced by adults. Infants may gradually convert from the structural to the dreaming form of REM sleep. This would help to explain the low incidence of SIDS during the first month.

It is also known that the neural "hardware" required for dreaming is not properly developed in a newborn. Hobson notes that many extensive connections in the cortex, as well as connections to and from the thalamus, which are important in relation to dreaming, are only fully developed in infants between one and three months of age (Hobson 1989, 78). This suggests that the machinery for dreaming (or the recollection of memory) is not properly developed in a newborn. This could go some way towards explaining the hiatus of SIDS during the first month after birth, as well as the comparatively low rate during the second month. Based on the exponential decay of memory, one would expect the incidence of SIDS during the second month after birth to be much higher than during the third month.

One should also note that infants do not really settle into a definite sleep pattern during the first month or two, and this may have some bearing on the low incidence of SIDS in the first two months after birth. In the first few weeks the average longest period of sleep is approximately four hours, and this steadily increases to about 8.5 hours at sixteen weeks of age (Parmelee, Wenner, and Schultz 1964).

Another important point about the age-at-death distribution is that it is the same for a premature infant and a full-term infant (Grether and Schulman 1989). This suggests that SIDS is not precisely linked with maturity, since a preterm baby is obviously less mature than a full-term baby. In other words, the risk of SIDS is a function not of the actual age of an infant, but of the age since birth. SIDS theories based on immaturity have trouble explaining this fact (Hillman 1991). In the FMD theory the clock starts at birth, since this is when dream development and dreaming begin, and this offers a natural explanation of why preterm and full-term infants sit on the same age-at-death distribution curve.

Risk Factors

Certain risk factors have been identified in relation to SIDS. The problem is that they are usually based on epidemiological data gathered from surveys and questionnaires, which are susceptible to error, bias, and misinterpretation. Risk factors may also be noncausal, that is, associated with something else that is more directly linked to the cause of death, or the trigger mechanism for SIDS. Risk factors are also intimately related to each other, and it is extremely difficult to factor out their interdependence. For instance, the risk factors associated with maternal smoking, low birth weight, and low socioeconomic status (see below) are clearly related to each other. Mothers from a lower socioeconomic group tend to smoke more often during pregnancy and have lower-birth-weight babies. Some other risk factors may have no direct impact on the pathogenesis of SIDS. For example, the risk of SIDS is reportedly less if a mother attends antenatal or postnatal clinics (Kohlendorfer et al. 1997). This is probably true only because most mothers who attend clinics are better informed about the dangers of SIDS and what precautions may or should be taken.

What is quite unusual about SIDS is that there are so many risk factors, some of which appear to be of a social nature only. This is quite peculiar. Some people have suggested, on the basis of this, that the cause of SIDS may be multifactorial, that is, SIDS may have lots of different causes. I believe however that this view is an inadequate explanation. One would imagine that there would be many more pathological signs associated with a SIDS death if it had multiple causes. The existence of social risk factors is indicative of the fact that something unusual underlies the cause of death in SIDS. SIDS may well have no simple medical explanation. This notion agrees with the FMD hypothesis, which asserts that an infant's mental state is of paramount importance. As we shall see below when we look at each of the main risk factors in turn, not only is the FMD hypothesis consistent with all of the risk factors, but in most cases there is no other completely satisfactory explanation.

The Prone Sleeping Position

The prone sleeping position (on the stomach, or facedown; see figure 6.2) is associated with a much higher risk of SIDS. It has been estimated that the risk may be anywhere from three times to ten times higher if a baby is placed to sleep in the prone position compared to other sleeping positions. All international SIDS foundations now recommend that babies should be placed to sleep in the supine position (on the back, or faceup; see Figure 6.4) or on the side (Figure 6.5). This recommendation is based on epidemiological research and is a well-established risk factor, although aside from my theory there is no proper medical or theoretical explanation as to why the risk should be so much higher if a baby is put to sleep in the prone position, yet still vulnerable in the supine position.

This risk factor was discovered when a link was noticed between a sudden increase in SIDS and a change in the sleeping position for babies in the Netherlands (Engleberts and de Jonge 1990; de Jonge et al. 1989). Prior to 1972, the

Dutch would mainly sleep their babies either on their back or on their side. In 1971 two influential scientists presented a paper at a conference in Vienna asserting that the prone sleeping position was (psychologically) healthier for infants. This convinced pediatricians and the media in the Netherlands to promote the prone sleeping position. Over the next fifteen years, from 1973 to 1987, the incidence of SIDS jumped tenfold from around 0.1 to more than one death per thousand births.

The risk associated with the prone sleeping position has now been confirmed by many other studies (reviewed in Dwyer and Ponsonby 1996). Since 1992, most Western countries have taken part in the so-called back-to-sleep campaign, which promotes the placement of infants in a nonprone sleeping position, either on the back or on the side. Based on this alone, the incidence of SIDS in most of these countries has halved, or more than halved. This decrease in the incidence of SIDS is not due to any changes in the diagnostic criteria used by pathologists during autopsy (Byard 1997). The incidence in the Netherlands has since fallen back to about 0.26 deaths per thousand live births (L'Hoir et al. 1998c). The discrepancy with the original lower rate may have something to do with the fact that prior to 1975 a disproportionate number of deaths were recorded as "accidental suffocation in bed" (Engleberts and de Jonge 1990). In Norway, Denmark, and Sweden, the incidence of SIDS has fallen from 2.3, 1.6, and 1.0 respectively to fewer than 0.6 deaths per thousand live births since the start of the back-to-sleep campaign. In Australia and the United States the rate has gone from around two to one in a thousand.

Most infants are now placed to sleep on their back or side, with around 15 percent still placed in the prone position. It is now possible to determine the risk associated with the two "safe" sleeping positions in relation to each other. A recent study suggests that the side sleeping position is associated with a relative risk factor of around 2.0 compared to the supine position (Scragg and Mitchell 1998). Other studies, however, have not found any significant difference between the side and supine positions (Ponsonby et al. 1995). It is not clear whether the main reason the side sleeping position may be of higher risk is that infants are more likely to roll over into the prone position from the side position (Fleming et al. 1996). This possibility has been suggested because some infants who were claimed to have been placed to sleep on their sides were found in the prone position at death.

In Sweden, it is estimated that around 15 percent of infants now sleep prone, compared to 72 percent prior to the back-to-sleep campaign (Lindgren et al. 1998). In the United States, before the back-to-sleep campaign, some 70 percent of infants were placed to sleep in the prone position in 1992, compared to about 24 percent by 1996, after the campaign (Willinger et al. 1998). As a result of these changes, the incidence of SIDS has about halved in both Sweden and the United States. With a little mathematics one can determine that the risk associated with the prone position compared to the back and side positions is around 3.3 using the Swedish data, and 4.5 using the American data. Assuming that half of the infants in the nonprone positions slept on the side and half on the back, and that the side position has twice the risk of the back position, this implies (with a little more mathematics) that the prone position is around five and seven times higher in risk than the supine position, respectively, for the two data sets.

In Asian countries like Hong Kong and Japan and in Finland and the Netherlands, the incidence of SIDS is traditionally much lower than in other industrialized countries, like the United Kingdom, Australia, New Zealand, and the United States (Davies 1985). This difference can be largely attributed to cultural practices, since in countries with a low incidence, the usual practice is to sleep infants either on their back or on their side. The lower incidence of SIDS in Hong Kong and the associated sleeping positions were instrumental in the discovery of the higher risk associated with the prone sleeping position (Lee et al. 1989). Other factors like climatic variation (see below) are also important in understanding the variations in SIDS among different countries.

It is particularly important to note that babies also die of SIDS in the supine position. This is known because some SIDS infants were found dead in the supine position at ages before they were able to roll over onto their backs.

It is quite remarkable that something as seemingly trivial as the sleeping position of a baby can have such a dramatic effect on the incidence of SIDS. The FMD theory is capable of explaining this (see below). Other theories are stretched to explain this extraordinary fact, and often cannot explain why infants also die in the supine position. It was popular to believe at one time that babies were prone to suffocation when they slept in the prone position, or that sleeping prone might interfere with the effectiveness of gasping. Such ideas cannot explain why babies die in the supine position, and there are no typical signs of suffocation in SIDS infants.

Another theory asserts that SIDS may be caused by the restriction of blood flow in the neck, and in particular into the brain stem of an infant, where the important cardiorespiratory, chemoreceptive, and arousal functions are located. This may interfere with some of these important brain-stem functions and upset the ability of the infant to respond to life-threatening apneic and hypoxic situations. Researchers have found that the body and head position do in fact influence the amount of blood that flows in the vertebral artery, which carries blood through the neck to the brain (Deeg, Bettendorf, and Alderath 1998; Pamphlett, Raisanen, and Kum-Jew 1999). The prone position can restrict this blood flow by pinching the artery at the junction where the spine and the skull meet. But I do not believe that this restricted blood flow can explain SIDS (although it may be a contributing factor), simply because infants also die in the supine position. This theory is also unable to explain why SIDS infants stop breathing in the first place, not to mention the many other facts about SIDS.

Others have proposed that SIDS is caused by excessive exposure to carbon dioxide (CO_2) (Corbyn 1993). The idea is that when an infant is either covered or is sleeping prone, this increases its exposure to CO_2 accumulated from expired air. This theory asserts that the cause of death is related to some sort of CO_2 poisoning or accidental suffocation. Small quantities (around 5 percent) of CO_2 in inspired air, however, are known to stimulate respiration. While CO_2 is not toxic in itself, it is thought that large amounts can lead to mental confusion and possibly cause death by suffocation. Although this is an interesting theory, it cannot account for some of the other features of SIDS, such as the age-at-death distribution, the deaths of supine infants, and the lack of pathological signs associated with this cause of death.

Another explanation offered for the higher risk associated with the prone sleeping position (and the climatic variation in SIDS, to be discussed in the next section) is based on "thermal stress" (Beal and Byard 1994; Guntheroth and Spiers 2001; Nelson, Taylor, and Weatheral 1989; Sawczenko and Fleming 1996). It is known that infants release a considerable portion of heat through their stomachs and faces (see also the discussion in the next section). Placing a baby to sleep in the prone position may increase thermal stress, the infant may overheat, and this may subsequently lead to apnea. One presumes that it may also disrupt the infant's arousal-from-sleep mechanisms. How these things happen is not entirely clear, and in any case this theory does not explain why most deaths occur during sleep, other than to suggest that a baby's defensive mechanisms are less reliable then. One might nevertheless expect more deaths in the awake state. Thermal stress theories, on their own, also do not explain the other data on SIDS such as the age at death.

In the FMD theory, the role of the prone position in relation to SIDS is quite clear. When a baby is put to sleep in the prone position, it generally tucks its legs up into its stomach, with its arms alongside its head (see figure 6.2). It assumes an almost fetal position (figure 6.3), and this could encourage fetal dreams. In the supine sleeping position, the baby generally has its arms and legs spread out. In deriving my theory I implicitly assumed, based on evidence from lucid-dreaming research, that during REM sleep the mind can influence the body. The converse may also be true, namely that the environment of an infant (which includes its body position) can influence dream content. We have all experienced dreams in which we imagine ourselves to be in an extremely cold climate, only to find on waking that the blankets have fallen off the bed. We also often feel that we need to urinate in our dreams, because there is a real physiological need to do so.

In the prone position, the infant is generally in the fetal position and this may more readily incite fetal dreams, which may potentially be more dangerous. The prone sleeping position may also invite the recall of fetal memory because a baby has its face pushed up hard against the mattress, possibly reminding it of being back in the womb, where the mother's organs were squashed up against its face, particularly during the latter part of pregnancy. In the prone position, the baby is also in a warmer environment because it is breathing its own exhaled air and because its temperature is harder to regulate. This may also remind the infant of being back in the warm womb. In the FMD theory, babies can also die in the supine position if they have a fatal fetal

Figure 6.2. An illustration of an infant sleeping in the prone (facedown) position. The infant is seen to assume an almost fetal position (see figure 6.3), with its legs tucked up into its stomach and its arms alongside its head. In the FMD theory this sleeping position may remind the infant of being back in the womb and increase the likelihood of fetal dreams and SIDS. *(Sketch by Anita Littlewood.)*

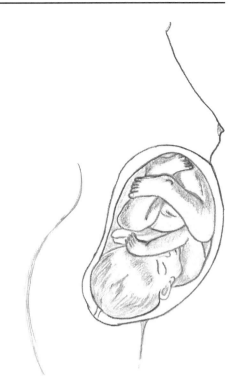

Figure 6.3. An illustration showing the approximate body position of a fetus in the womb. *(Sketch by Anita Littlewood.)*

dream, but there is a greater chance of an infant having a fetal dream in the prone position than in the supine position. The side sleeping position is arguably more dangerous than the supine position because the infant can more readily roll over into the prone position from the side position, and there are more characteristics of the side sleeping position that correlate with the fetal position (see figures 6.3 and 6.5).

A study of SIDS infants in Norway, Sweden, and Denmark found that the prone and side sleeping positions are of even higher risk for younger infants, from three to six months of age, compared to older infants, from six to twelve months (Oyen et al. 1997). The risk associated with the prone sleeping position increases further for infants who are premature or of low birth weight. This means that younger and preterm infants are more vulnerable in the prone sleeping position, a finding which is consistent with the FMD hypothesis since the prone position is more likely to stimulate the recall of fetal memory for a younger or preterm infant, in which fetal memory is expected to be stronger and REM sleep greater than in other groups.

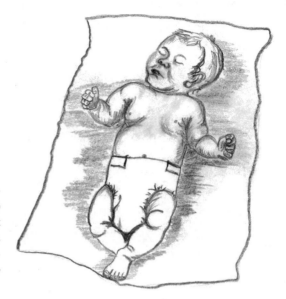

Figure 6.4. An illustration of an infant sleeping in the supine (faceup) position. In this position the infant generally has its arms and legs spread out, and there is little resemblance to the fetal position. *(Sketch by Anita Littlewood.)*

Figure 6.5. An illustration of an infant sleeping in the side position. This position is seen to have some similarity to the fetal position. *(Sketch by Anita Littlewood.)*

It has been suggested that the supine sleeping position is safer than the prone position because it is associated with more awakenings during sleep (Goto et al. 1999). This is presumed to have a preventive effect. "Bed sharing," however (see discussion below), is also associated with more arousals during sleep but does not seem to provide any protection against SIDS; indeed, it is actually regarded as a risk factor (Mosko, Richard, and McKenna 1997). This suggests that the protective nature of the supine sleeping position is not associated with more arousals per se.

The Role of Climate and the Infant's Sleeping Environment

The incidence of SIDS is generally higher in countries with colder climates. Within any country, the incidence of SIDS is generally highest in the coldest regions (Mitchell et al. 1991; Nelson and Taylor 1988; Ponsonby et al. 1992). This climatic risk factor is evident when one notes that the incidence of SIDS in Hobart (Tasmania), in the south of Australia (average annual temperature 10°C) is about ten times the rate in Darwin, in the north of Australia (average annual temperature 27°C) (Australian Bureau of Statistics 1990). Some of this difference can be attributed to cultural differences (even within Australia), but the main contribution to SIDS would have to come from something to do with the sleeping environment of the infants. Most researchers now believe that the way these infants are covered during sleep is the most important factor involved in the observed climatic variation of SIDS. There are of course some cold countries, such as Finland, Sweden, and Japan, where the incidence of SIDS is comparatively quite low. This may be attributed to different cultural practices, such as traditionally sleeping infants in a nonprone position, and to the fact that in countries like Sweden most homes are centrally heated, which is generally not the case in Australia. A heated room would mean that an infant does not need much bed covering; however, an overheated room may pose a risk through overheating.

The incidence of SIDS is generally highest in the coldest months of the year (Peterson, Sabotta, and Strickland 1988; Steele 1970). In Australia, New Zealand, and other Southern Hemisphere countries, the incidence of SIDS is two or three times higher in the winter months of June, July, and August compared to the summer months of December, January, and February (Mitchell et al. 1991). In the United States, Canada, and other Northern Hemisphere countries, the incidence of SIDS is highest in the winter months of January, February, and March. One should note, however, that the greater incidence of SIDS in the winter is not unique to SIDS (Spiers and Guntheroth 1997). The higher incidence in the coldest months originally led researchers to look for a bacterial or viral cause of SIDS, but this has largely been ruled out now. In the FMD theory the regional and seasonal variation in the incidence of SIDS may be attributed to the fact that during winter and in colder climates parents tend to cover a sleeping infant more extensively, and this, through the warmth and cushioning associated with some coverings, may remind the infant of being back in the womb.

One of the other main recommendations of most international SIDS foundations is to not cover a baby excessively. In particular it is suggested that one should not cover the baby's head during sleep (Henderson-Smart, Ponsonby, and Murphy 1998). There is evidence that covering with comforters (called "duvets" in England and Australia) and quilts increases the risk of SIDS (Fleming et al. 1996; Ponsonby et al. 1998). The standard view is that excessive covering, the prone position, and covering the head cause the baby to overheat and that thermal stress is an important factor in relation to SIDS (Sawczenko and Fleming 1996). In the womb, the fetus is surrounded by a heat bath of embryonic fluid at a temperature of around 37°C. Thermal stress may then remind the infant of being back in the womb. The FMD theory offers an explanation of why thermal stress may lead to apnea and SIDS. I would suggest that covering the baby's head could stimulate fetal dreams not only because of the associated heating and cushioning, but also because of the lack of light entering through the closed eyelids and the dulling of auditory senses, conditions that might be associated with memories of the womb. Comforters have a tendency to smother, which may remind the infant of its time in the womb.

I originally attempted to explain the link between SIDS and colder climatic conditions through the idea of wrapping, suggesting that tight wrapping might remind an infant of being back in the womb, where it was surrounded by tightly packed organs (Christos 1995a). This notion is probably not accurate because in the womb the baby is surrounded by embryonic fluid, which would tend to redistribute and relieve any pressure from the surrounding organs. I now believe that a smothering or covered feeling may be more in line with what a fetus may be experiencing in the womb. The epidemiological data actually suggests that tucking a baby in tight has a protective effect in relation to SIDS, particularly if it is placed to sleep in the supine position (Wilson et al. 1994). This probably prevents the baby from rolling over or curling up into a fetal position. Tight covering would also prevent the baby from sliding under the bed covers.

There is some indication in the literature that the prone position combined with other factors carries a greater risk than the product of the risk factors themselves. In other words, two risk factors seem to have a bootstrapping effect on each other. We saw earlier that the prone sleeping position was an even greater risk factor for a young and premature infant. In the same way, it is known that there is an increased risk associated with the prone position in the winter as opposed to the summer (Mitchell et al. 1999). I would argue that in the FMD theory the prone position and covering could work together to further increase the possibility of fetal dreams.

Thermal stress theories are unable to explain why more infants do not die in the hottest months of the year, when they also seem to be thermally stressed. Such theories generally presume that overheating results only when a baby is excessively covered or is overdressed during the winter. One suggestion is that the infant's cooling system is more effective during the summer months because sweating is intact, but by the same token sweating is also regarded as a risk factor for SIDS (Guntheroth and Spiers 2001). Sweating during winter is viewed as a sign of overheating. The difference between sweating in summer and in winter may have something to do with the fact that sweating in summer is effective because of evaporation, whereas sweating during winter, particularly under excessive bed clothes, is ineffective. If sweating (and damp hair) are indeed risk factors, the FMD theory would suggest that a hot and wet environment may help to stimulate fetal dreams.

More recently, some researchers have proposed that overheating of an infant's head, or brain in particular, may be the most important factor involved with this higher risk. One could argue that the baby would be more liable to overheat when it is placed in the prone position, with its head covered. It is known that when a baby is excessively covered, such as during winter, the head becomes the main location of heat loss. George du Boulay has proposed an interesting theory relating to this idea, arguing that the brain avoids overheating by cooling via the blood in the arteries of the face and neck (reported in *New Scientist*, January 1999). Using a tiny thermometer, he found that the temperature of the blood in the arteries going into the brain was 1.5°C cooler than blood flowing out of the brain. The idea is that when a baby is sleeping prone and/or is excessively covered, it is more susceptible to cranial overheating. There is evidence that an infant's head is subject to a higher temperature when it is sleeping prone (Oriot et al. 1998). This theory does not however explain how hyperthermia actually leads to death, why this happens during sleep, why infants in the supine position also die, and the age-at-death distribution.

Another interesting finding is that the age-at-death distribution is not completely independent of the seasonality factor. For example, younger infants are more likely to die earlier in the winter months (Douglas, Helms, and Jolliffe 1998). In the United Kingdom this means that babies less than four months of age are more likely to die in January, while older babies are more likely to die in February. This may be so because younger infants may be covered up more when the winter starts because of their size and perceived vulnerability to the cold.

The Socioeconomic Factor

There is a clear link between the incidence of SIDS and the socioeconomic status of families (Hoffman et al. 1988; Shrivastava, Davis, and Davies 1997). Infants from poorer families are generally at higher risk of SIDS than infants from families with greater means, but this is also true for most other childhood and infant diseases. One way it may influence SIDS is that poorer households are generally not well heated and parents may need to use more bed covers on their babies. This may then stimulate fetal dreams, because of either the excessive covering or the local warmth.

As noted earlier, since the introduction of the prone sleeping position, the incidence of SIDS has almost halved in most Western countries. This has introduced another factor into the association of SIDS with socioeconomic status, because people from poorer backgrounds tend to be less educated about such findings and tend to ignore the warnings about sleeping position. In one study, it was found that in three inner-city hospitals in the District of Columbia that predominantly serve poor families, up to 40 percent of mothers placed their babies to sleep in the prone position, compared to 15 percent for the national average (Brenner et al. 1998; Willinger et al. 1998). There is also evidence suggesting that people from a lower socioeconomic background tend to ignore the warning about smoking during pregnancy, which is regarded as a risk factor.

Maternal Smoking during Pregnancy

Maternal cigarette smoking is purported to be a risk factor associated with SIDS (Bultreys 1990; Hoffman et al. 1988; Malloy et al. 1988). Taking into account some other confounding factors, current estimates suggest that maternal smoking could double the risk of SIDS (Anderson and Cook 1997; Golding 1997). This risk factor is significant because 25 percent of all mothers smoke during pregnancy in the United States (Slotkin 1998). Evidence suggests that the risk is dose dependent, that is, higher the more cigarettes a mother smokes (Macdorman et al. 1997).

The general consensus is that smoking during pregnancy is the most significant contributor to this risk factor, although it is difficult to separate the effects from those of smoking after birth, because mothers who smoke during pregnancy tend to smoke after birth. One study measured the amount of urinary cotinine in infants, which is an indicator of the amount of nicotine exposure (Dwyer, Ponsonby, and Couper 1999). If a mother did not smoke in the same room as the infant, there was a reduced level of cotinine in the infant's urine, but this was not associated with a reduction in the risk of SIDS. Furthermore, there does not seem to be an associated risk with other members of the family smoking, and although passive smoking may be linked to significant structural changes in the airways of infants, this is unlikely to be a primary cause in relation to SIDS, since SIDS infants are generally not found to have an obstructed airway passage at autopsy (Elliot, Vullermin, and Robinson 1998).

Many researchers believe that smoking and SIDS are directly linked, but the situation is not that clear, as maternal smoking is related to other risk factors and is generally implicated in other perinatal deaths, cardiorespiratory diseases, and hypertension (high blood pressure). Even if there was a fundamental link between SIDS and nicotine exposure, this would not explain the etiology of SIDS because infants who are not exposed to nicotine also die.

The favored notion is that smoking, particularly during some critical period during pregnancy, may interfere in some way with the normal physical and neural development of a fetus or infant, that diminishing its ability to respond to life-threatening situations later. Smoking may retard the development of the infant's respiratory system, cardiovascular system, chemoreceptor/reflex systems, or arousal systems. Nicotine is actually used in some parts of the brain as a neuroreceptor, and one theory suggests that exposure of nicotine to a fetus or infant may alter the nicotinic receptors in brain-stem nuclei that are related to cardiorespiratory function and arousal (Nachmanoff et al. 1998). A study with rats found that fetal exposure to nicotine impairs the rat's ability to respond to hypoxic (lack of oxygen) events during apnea (Fewell and Smith 1998). Brain-stem gliosis (see the section below on peculiar medical findings), which is linked to smoking, has also been observed in some SIDS infants.

Smoking is however, also strongly linked to the socioeconomic risk factor, and its associated risk factors. Poorer people tend to ignore health warnings about smoking and are more likely to ignore warnings about smoking during pregnancy (Jorm et al. 1999). Socioeconomic status is a risk factor for the reasons outlined previously. There is also an increased risk associated with mothers who use illicit drugs (other than small amounts of alcohol) during pregnancy (Hoffman et al. 1988). An exception to this is cocaine, which does not seem to bring a higher risk for SIDS if used by a mother during pregnancy (Hillman 1991; Valdes-Dapena 1991). Some results even seem to suggest (taken at face value) that cocaine use is protective against SIDS. This may however simply reflect the fact that cocaine is generally a high-class (expensive) drug, which is abused less often by people from a lower socioeconomic class. These observations add credence to the notion that a considerable part of the risk associated with drugs, including nicotine, may have a socioeconomic basis.

I believe that smoking is not such an important risk factor (in its own right) as is suggested in the literature, because there are many confounding factors, and the various risk factors are interrelated. The main influence of smoking seems to be during pregnancy. This may interfere with the normal neural development of an infant's important brain-stem systems. The FMD theory makes no formal connection with maternal smoking other than to take on board that nicotine exposure, particularly during some critical period, may impede the development of these vital systems.

Ultimately the only way to know for certain what the true impact of smoking is on SIDS is to see what happens to the incidence of SIDS when most mothers stop smoking during pregnancy.

Some Other Risk Factors

There are other risk factors associated with SIDS, but many of these are considered to be of secondary importance or are related to some of the other more important risk factors already discussed. Most of these risk factors, including maternal smoking, were identified in the pioneering work of Hoffman and his colleagues (1988).

SIDS is more frequent in premature (preterm) babies, babies with a lower birth weight, infants belonging to young and single mothers, black infants (compared to white infants), infants from crowded homes, infants whose mother's education is less than twelve years, and infants who have been exposed to venereal disease or urinary-tract infections. Premature babies have more REM sleep, and their life-preserving neural circuits are probably not fully developed, factors that may affect SIDS in the FMD theory. All of these risk factors are also obviously related to the socioeconomic risk factor and factors associated with it, like smoking, poor heating, excessive covering, and ignoring warnings about the prone position.

Male babies are more susceptible to SIDS than female babies, especially in the first year of life (Hoffman et al. 1988). Approximately 60 percent of all SIDS victims are males, and 40 percent are females. Another study suggests that the difference may not be as high as this, giving figures of 52 and 48 percent respectively (Valdes-Dapena 1991). Researchers have verified that this difference cannot be attributed to the male babies being exposed to more of the previously identified risk factors (Mitchell and Stewart 1997). The higher risk for male babies has led to one theory involving testosterone, but this theory cannot account for any of the other features of SIDS. The discrepancy between male and female babies might result if males matured at a slower rate than females in some important way. One should note that male babies generally do worse than female babies in almost all other respects; for example, more die from lung disease and congenital defects.

The risk of SIDS is reduced if an infant sleeps in the same room as adults but is increased if the infant actually shares a bed with adults. This seemingly trivial fact fits in nicely with the FMD theory, since I would suggest that bed sharing may add a heartbeat to the infant's sleeping environment, which may remind it of being back in the womb, whereas sleeping in the same room but not the same bed may act as a reminder to the infant that it has been born and so would be protective. Bed sharing may also be linked to SIDS through the associated local warmth and covering factors. It has been suggested that when a baby sleeps in the same bed as its mother, there is a greater tendency for its head to be covered, leading to possible thermal stress (Guntheroth and Spiers 2001).

SIDS is mildly linked to certain infant medical conditions, many of which were discovered after going over medical records of newborns in hospital nurseries (Hoffman et al. 1988). Some reported neonatal risk factors include tachycardia (accelerated heart rate), hypothermia (lower than normal body temperature), tachypnea (accelerated breathing), pallor (pale color), cyanosis (discoloring of skin due to lack of oxygen in the blood), poor feeding, vomiting, irritability, fever, respiratory distress, lethargy, an abnormal cry (whatever that means?), and possibly newborn apnea (Hoffman et al. 1988).

SIDS is more frequent on weekends than weekdays (Williams et al. 1997). This mysterious observation may be explained by the fact that there is generally less sharing of the bedroom with an adult during a weekend. This increase may also be related to the fact that infants and parents tend to sleep in for longer over weekends, and extended sleep generally means more and longer periods of REM sleep. There is a reported higher risk associated with SIDS if an infant sleeps away from home (Schluter et al. 1998). The authors try to explain this in terms of a decrease in breast feeding and room sharing, but breast feeding is no longer recognized as an important factor in relation to SIDS (see below). I would suggest that perhaps when an infant is away from home it may also be sleeping longer, especially in the morning (when most REM sleep occurs). There is also a reported increase in SIDS for infants whose mothers have moved house after birth and for infants who have lived in numerous dwellings. These risk factors may be simply associated with the socioeconomic factor, since poorer families would have a higher tendency to rent and to move house.

Soft bedding and mattresses are considered to be risk factors (Mitchell, Scragg, and Clemerts 1996; Scheers, Dayton, and Kemp 1998). Sheepskin bedding is a risk factor if an infant is sleeping prone (Mitchell et al. 1998). In another study, non-SIDS infants were found to have slept on a firm mattress more often than SIDS infants (L'Hoir et al. 1998a). I would suggest that the cushioned and possibly warmer local environment offered by soft bedding (particularly if a baby's face is pushed into a soft mattress cover and it is sleeping prone) may enhance the prospects of fetal dreams.

With the clear reduction of SIDS as a result of asking parents to avoid the prone sleeping position for their babies, there is now an educational and apathy risk factor. Some people may not be aware of the risk associated with the prone position, or they may not believe the advice offered to them and thus ignore it.

Some Peculiar Medical Findings

Researchers have noticed certain abnormalities in SIDS infants as a group, such as a deficiency or alteration of some neurotransmitters or neuroreceptors, delayed maturation of the brain stem, brain-stem gliosis, and certain features in sleep organization and cardiorespiratory function. These findings are not observed in all SIDS cases, however, so they cannot be used as markers to identify SIDS at postmortem or to determine which infants are most at risk.

Although the cause of death in SIDS is not known, it is widely believed that the primary cause is irreversible respiratory cessation (Naeye 1980). Pathology in SIDS is not inconsistent with respiratory failure. This includes the observation of petechiae, or small broken blood vessels, found on surfaces of the lungs and heart of SIDS infants, and the increased retention of brown fat (Beckwith 1988; Valdes-Dapena 1983). The petechiae are observed in approximately 80 percent of all SIDS cases (Krous 1988). As mentioned earlier, there are also some cases that were originally misdiagnosed as SIDS but have since been reclassified as suffocation murders. This suggests that it may not be so easy to distinguish between SIDS and mechanical asphyxia (the insufficient intake of oxygen) (Hata et al. 1997). These findings are consistent with the FMD hypothesis, which affirms that the primary cause of death is respiratory cessation.

At postmortem, SIDS infants are observed to have elevated levels of fetal hemoglobin, although one study finds this is not the case (Giulian, Gilbert, and Moss 1987; Perry, Vargascuba, and Vertes 1997; Zielke et al. 1989). When an infant is born, its fetal hemoglobin is slowly replaced by adult hemoglobin. This process is normally completed within the first six months after birth. SIDS infants have been found to have more fetal hemoglobin than they should have at their particular age. As these higher levels are not observed in every SIDS infant, they may not be directly linked to SIDS and so cannot be used as a marker to identify infants at risk. Fetal hemoglobin carries more oxygen than adult hemoglobin, so this suggests that SIDS infants may have experienced repeated episodes of hypoxia (or oxygen deficiency), which in turn may have been caused by fetal dreams.

Some SIDS infants have been observed to have high levels of brain-stem gliosis, the proliferation of glial cells, whose main function is to nourish and support neurons and remove debris (Becker 1990; Kinney and Filiano 1988; Goyco and Beckerman 1990; Kinney et al. 1991). The amount of gliosis is also thought to be positively linked to the amount of nicotine exposure during pregnancy (Storm, Nylander, and Saugstad 1999). Other SIDS infants have altered levels of certain neurotransmitters or neuroreceptors (Obonai et al. 1998; Panigrahy et al. 1997; Kinney et al. 1998). These features, however, are not observed in all SIDS cases, so they cannot be the cause of death, although they can certainly contribute to it.

Differences in sleep organization, heart rate, and respiratory patterns have been observed in SIDS infants as a group. Data from a number of hospitals in Los Angeles suggest there may be some very subtle differences between SIDS and normal infants. For instance, SIDS infants younger than one month showed significantly more units of REM sleep across the night compared to normal infants, and SIDS infants for all ages studied seemed to have a higher proportion of REM sleep, particularly at the end of their main sleeping period (Schechtman et al. 1992). SIDS infants also showed reduced motility (or spontaneous movement) during REM sleep compared to control infants. These findings indirectly support the FMD hypothesis, since the probability of fetal dreams increases with increased amounts of REM sleep and such dreams might be expected to involve less impetuous movement. Another sleep study found that the victims of SIDS had a higher average heart rate than control infants (Schechtman et al. 1988). Even more interesting is the fact that, for SIDS infants older than one month, this increased heart rate is reported for REM sleep only. In terms of the FMD hypothesis, if an infant is dreaming that it is a fetus, it may show other signs characteristic of a fetus, such as a higher heart rate. Yet another study found that infants who were vulnerable to life-threatening ventilatory problems had REM sleep abnormalities such as a significant increase in the amount of REM sleep from 2 to 5 a.m., which is regarded as the most critical period for SIDS (Cornwell, Feigenbaum, and Kim 1998). SIDS infants also seemed to have subtle differences in their breath-to-breath respiratory patterns during sleep, particularly for low respiratory rates (Schechtman et al. 1996).

Another study found that around half of all SIDS infants had a condition called "long QT interval," a sort of arrhythmia of the heart (Schwartz et al. 1998). Peter Schwartz has suggested (as reported in *New Scientist,* April 1999) that we should screen babies for this condition. The suggestion has met with considerable opposition, because it is too expensive to screen all babies, and 98 percent of infants with long QT interval do not die of SIDS (Southall 1999). The most recent results from Schwartz and his coworkers have not been confirmed, and an earlier study found no significant difference between SIDS infants and age-matched controls in relation to QT intervals (Southall et al. 1986). It is not clear if Schwartz believes that this heart abnormality can cause SIDS. It certainly cannot account for all SIDS cases, not to mention explaining many of the other features of SIDS.

It has been reported that SIDS infants dying on cardiorespiratory monitors, while in hospital and at home, show progressive bradycardia (slowing of heart rate below sixty beats per minute [bpm]) as they die (Poets et al. 1999; Ledwidge, Fox, and Mathews 1998). An infant's heart rate is normally around 120 bpm, while a fetus has a rate of 120 to 160 bpm. This finding would seem, at first sight, to contradict the FMD theory, because one might imagine that if an infant was having fetal dreams it would display a higher heart rate. As we get older however, our heart rate slows naturally, to about 70 bpm by the age of eighteen years, for example. A higher heart rate generally corresponds to a greater need for oxygen during our most rapid growth periods. Heart rate is thus a measure of the amount of oxygen required. One could argue that the observed bradycardia in SIDS infants occurs because there is less need for atmospheric oxygen by an infant dreaming of being a fetus.

Other Odds and Ends

It had been reported that the risk of SIDS is reduced if a baby is breast-fed (Hoffman et al. 1988). In 1992 the National SIDS Council of Australia began promoting breast-feeding as a protective measure for SIDS. Recent research, however, suggests that this is not such an important risk reduction factor, although there is evidence that non-SIDS infants are more often breast-fed than SIDS infants (Fleming et al. 1996; L'Hoir et al. 1998a). There may well be some other reason for this correlation. Hillman has questioned the relevance of breast-feeding in relation to SIDS on the grounds that, even though the percentage of breast-fed babies increased over the 1980s, the incidence of SIDS did not decline over the same period (Hillman 1991). The Australian SIDS Foundation no longer recommends breast-feeding as a (major) risk reduction measure for SIDS, although the practice probably does have other benefits for infants. As noted previously, there are some reported cases in which an infant has succumbed to SIDS immediately after breast-feeding, in some cases while still in its mother's arms. Breast-feeding did not seem to help prevent SIDS in these cases. This turnabout in relation to breast-feeding is supportive of the FMD theory, as it is difficult to come up with a reason why breast-feeding per se should be so important. (See however the discussion at the end of this chapter regarding pacifiers and thumb sucking.)

As noted earlier, the highest incidence of SIDS occurs in the two- to four-month age group, a time when babies are immunized, but there is no evidence to suggest that SIDS is linked to immunization. It is not caused by vomiting, choking, or suffocation, and it is not contagious or caused by an infectious disease or a minor illness such as a cold. SIDS is not brought on by an allergic hypersensitive reaction of the body to a foreign protein or drug (Hagan et al. 1998). Moderate alcohol use during pregnancy is not a risk factor in relation to SIDS. This is somewhat surprising since most other drugs, such as nicotine, have a negative effect. SIDS infants do not show any signs of immaturity in the development of their diaphragms (the main respiratory muscle) or the phrenic motor nerves that operate the diaphragm (Weis et al. 1998)

Originally it was thought that certain families were more susceptible to

SIDS, since in a number of cases more than one infant in the same family died from it. This led to the suggestion that there may be a genetic link to SIDS, and numerous studies were conducted on SIDS siblings (brothers and sisters of SIDS infants) that were still alive to try to find the cause of SIDS. It should be noted that siblings tend to experience the same environmental conditions, so the greater incidence of SIDS in certain families may have to do not with genetics but with the common environment. Current thinking is that there is no direct genetic factor involved in SIDS. It is known, for example, that SIDS rates are not higher in identical twins compared to nonidentical twins (Peterson, Chin, and Fisher 1980). Numerous papers persist in reporting a strong family link, but as noted above, this may be more simply related to the fact that infants from the same family experience similar environmental conditions, and as we have seen, environmental factors (like sleeping position) are vitally important in SIDS.

One of the reasons why people thought that SIDS had a strong link with families was that it was reported in an influential paper by Alfred Steinschneider, which also suggested a strong link between sleep apnea and SIDS (Steinschneider 1972). Steinschneider studied families with multiple cases of SIDS, which happened to also have histories of severe or prolonged sleep apnea. The problem was that some of the infants Steinschneider studied had been murdered (Firstman and Talan 1997). One of the families had lost five children, all presumed to be SIDS victims. The mother later confessed to police that she had murdered all five infants. A subsequent revision of the theory that SIDS runs in families has prompted a reexamination of other cases in which families have lost more than one infant to SIDS. Another mother in the United States has since been charged with the murder of six of her children, all previously diagnosed as SIDS. Firstman and Talan conclude that as many as ten percent of SIDS deaths may actually be infanticides. This may be an overestimate.

Another idea inspired by Steinschneider's work was that SIDS was related to apparent life-threatening events (ALTEs), but this too seems to be under skeptical scrutiny. As noted earlier, subsequent studies have concluded that prolonged apnea and ALTEs in infants are not predictive of SIDS, and there is no direct proof that ALTEs were aborted SIDS cases or near-miss SIDS (Southall et al. 1982). Very few infants experiencing ALTEs have succumbed to SIDS. Consequently hundreds of studies involving ALTE infants may have no relevance to SIDS and should probably be discarded. Since it now appears that there is no genetic link, studies that use SIDS siblings should probably be discarded as well, although, as noted above, siblings may share a common environmental factor.

The presumed connection between sleep apnea and SIDS led to a vast amount of research on developing monitoring devices to detect ALTE and to alert parents when a baby stopped breathing. Even if SIDS is linked to this form of apnea, such monitoring is of limited use because it is expensive and consequently does not necessarily target the infants who are most at risk, those from a lower socioeconomic background. Moreover, a number of SIDS infants have died while they were being monitored, so it is not clear that monitors work in any case (Ledwidge, Fox, and Mathews 1998; Poets et al. 1999).

TESTING THE FMD HYPOTHESIS

In science, one can never actually prove that a theory is correct. All one can do is devise experiments to test the theory further, or try to prove it is false or deficient. The theory would then need to be revised or updated, and if this was not possible, it would have to be discarded. One of the problems with the FMD hypothesis is that it is difficult to test, because it asserts that the cause of SIDS lies in the mind of an infant, a region that is very difficult to probe. This does not mean that one should dismiss the theory, although it may explain why it has been so difficult to find a satisfactory explanation for SIDS up until now.

The FMD hypothesis can of course be tested against all the known facts about SIDS, as I have done. As the hypothesis is consistent with these facts, it is a plausible theory for the etiology of SIDS. The FMD theory not only explains what the cause of death is (cessation of breathing), but also explains what triggers this death (recollection of fetal breathing pathways), when it happens (during REM sleep), and why all of the chemoreceptor and arousal mechanisms fail simultaneously (they are not needed). This situation should be contrasted with other theories, which are generally able to account for only one or two of the known facts and do not address these other issues. This is probably why these theories are normally prefaced with the statement that "this may explain some cases of SIDS," and why many researchers assert that SIDS is due to multiple causes. The FMD theory may well be the only truly viable single-cause theory of SIDS.

One of the most obvious tests of the FMD theory is to see if SIDS does occur during REM sleep. It is not feasible to permanently monitor infants with an

EEG (electroencephalogram) until one does succumb to SIDS, to find out if it happens during REM sleep. A rough estimate suggests that one would have to monitor infants for about two hundred thousand "infant nights" to observe just one such event. This number comes from the fact that on average one in a thousand of all live births results in SIDS, and this death can occur anytime within a period extending roughly six months (approximately two hundred days). Monitoring is also unethical, in that if an infant was being monitored it should be woken up as soon as it became apparent that something was wrong. A SIDS death may of course occur coincidentally while an infant is being monitored for some other reason. This has actually happened (two references were given earlier), but as far as I know there is no indication whether it occurred during REM sleep.

A more realistic approach is to record the time of death in relation to the time since the infant last went to sleep. For a specific age, infants have a fairly definite pattern of alternating cycles of slow-wave (SW) and REM sleep following sleep onset, as shown in figure 5.1, but this pattern is different for different age groups, and there is some variation even within age groups. If one were to plot as a frequency histogram the number of SIDS deaths in a specific age group as a function of the time since the start of sleep, it may look something like what is shown in figure 6.6 (if the FMD theory is correct). One would then be able to correlate this with an infant's age-specific sleep pattern, as shown in the inset in figure 6.6.

There are two problems with this. First, the time of death is not always known. A dead infant is normally found in the morning, and may have been dead for some time. It is not a simple matter to infer the time of death at an autopsy, because there are hardly any pathological signs of the cause of death. Second, in the first few months, an infant's sleep pattern fluctuates quite dramatically, and does not really settle into a definite pattern until about three to four months of age (Hopperbrouwers et al. 1988; Ditlrichova 1966). There are also unusual occurrences, as was mentioned earlier, in which an infant may go directly into REM sleep at sleep onset, even though the usual practice is for an infant to first enter into SW sleep. I believe, however, that by looking closely at the lapsed time from when infants fell asleep until they died, particularly for infants in the age group of three to six months, one might be able to tell whether SIDS occurs during REM sleep. Such data could be gathered by the numerous SIDS foundations around the world.

Another way to test the FMD theory is to see if SIDS correlates with any

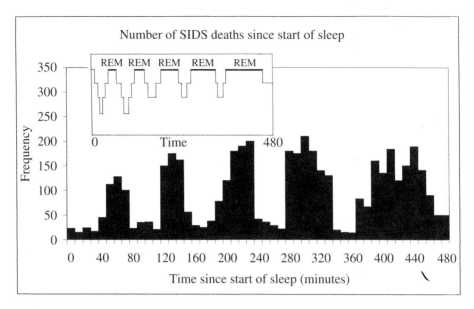

Figure 6.6. A hypothetical histogram data set showing the number of infants that died of SIDS as a function of the time since they started sleep. Here time is broken down into ten-minute intervals. One could try to correlate this data with the sleep pattern of infants for a specific age group to determine if the time of death is correlated with REM sleep. The inset shows a typical infant's sleep pattern.

other signs that are specific to a fetus, such as a higher average heart rate. Once again this would be difficult to establish for an individual infant because it is so difficult to capture that solitary moment, and, in any case, one would hardly believe something was true based on a single event. One could however look back at hospital data, as many researchers are currently doing, to derive information about SIDS infants as a group. Admittedly some SIDS infants have demonstrated progressive bradycardia (slowing of heart rate) just before death, but as noted earlier, this may have more to do with a diminished need for oxygen. It is interesting though that tachycardia (or an accelerated heart rate) in a neonate is regarded as a risk factor for SIDS (Hoffman et al. 1988). Could tachycardia help to stimulate fetal dreams?

Another idea is to use animals in some way. If SIDS is indeed linked to REM sleep as suggested by the FMD theory, then since almost all mammals also have REM sleep, they should theoretically also have SIDS. I am unaware if any animal deaths have been recorded as a SIDS equivalent. One could study animals (particularly those with a long gestation period) in detail during sleep and try to induce fetal dreams by making the environmental conditions as much as possible like those the animals might have experienced when they were in the womb. This might increase the chance of observing a SIDS animal death in the laboratory.

Another way to test my theory is to devise new preventive measures that may discourage infants from having fetal dreams, and see if this does influence SIDS in some way. Some ideas along these lines are considered in the next section on prevention. One might be able to test my theory by looking at the incidence of SIDS in babies who have been unintentionally exposed to fetal-like conditions. Babies who suffer from colic are often given toys or devices that make womblike sounds or have a recorded heartbeat. Is SIDS higher for these groups?

PREVENTION

If my explanation of the cause of SIDS is correct, it would imply that there is no miracle cure or vaccine for SIDS. Dreaming is an essential part of the development of an infant. I would certainly not recommend that an infant be deprived of REM sleep, because REM sleep clearly serves an important biological function. The best we can do is take preventive measures that may reduce the likelihood of a fetal dream.

My theory is in agreement with the main recommendations made by the international SIDS community. A forum convened by the National SIDS Council of Australia in Melbourne in March 1997, involving thirty-eight scientists, reviewed the risk factors associated with SIDS, gleaned from epidemiological studies (Henderson-Smart, Ponsonby, and Murphy 1998).

Its recommendations were as follows:

- Infants should not be placed in a prone sleeping position unless there is some other medical reason to do so. Sleep a baby on the back or side. The side position seems to be associated with a slightly higher risk compared to the supine position, so the supine position should be used as a first preference. There is of course a perceived danger that if infants are sick they may choke on their vomit if they are sleeping on their backs. A recent study, however, found that babies are more likely to vomit and choke when they

are placed to sleep in the prone position (Hunt, Fleming, and Golding 1997).
- Infants should not be covered excessively during sleep. Avoid comforters and quilts. It is recommended that a baby be placed to sleep in the supine position and tucked in tightly, with as few bed covers as possible. A tight tucking in prevents the baby from rolling over into the prone position or slipping under the covers. One should maintain a comfortable temperature in the room, preferably on the cool side, to ensure that a heavy covering is not required. It is also recommended that the baby's head not be covered.
- Mothers should not smoke or do anything else during pregnancy that might affect the normal development of the fetus, as this may impede the infant's ability to respond to life-threatening situations later.

There is another very important way in which we can all help to reduce the incidence of SIDS: informing friends and others about the dangers associated with the prone sleeping position and excess covering during sleep. International SIDS foundations are working hard on this, but we should all try to do more as a community. There is a tendency for people not to raise these issues with friends who have just had a baby, because SIDS is a rather frightening topic for young parents. A considerable decline in the incidence of SIDS has occurred since the foundations started the "back-to-sleep" campaign. In Australia the incidence of SIDS has declined by as much as 60 percent. It is known, however, that somewhere around 15 percent of mothers still sleep infants in the prone position, and many mothers may not be aware of the risks associated with covering and smoking. A recent news report in Australia suggested that up to 30-40 percent of mothers still smoke during pregnancy.

On the basis of the FMD theory, I would recommend that one try to ensure that the environment of an infant is as unlike the womb as possible. We may try to add things to the infant's sleeping environment that remind it that it has been born and is not in the womb. One idea may be to add a little noise to the infant's room, such as playing a radio. Noises may be important in relation to reducing SIDS, as sounds in the womb would have been muffled. A crunchy-sounding mattress may be helpful. As noted before, the risk of SIDS seems to be reduced if an infant sleeps in the same room as other adults but increases again if the infant sleeps in the same bed as the adults (Scragg et al. 1996). Maybe infants should sleep in the same room as their parents. Having air constantly moving past an infant's face while it is asleep may be helpful. This is something that would not have been experienced by a fetus in the womb.

Another idea is to give the infant a pacifier to suck while it is sleeping. This may remind the sleeping infant that it is now alive and not in the womb, and it may be more effective if the baby is breast-fed as well. One might reject this idea on the grounds that fetuses are regularly observed, by ultrasound, to be sucking their thumbs in the womb. Indeed thumb sucking is often listed as one of the things a fetus is able to do when it is seven months old. The epidemiological data is particularly interesting on this point, because two studies suggest that pacifiers are preventive toward SIDS, whereas another study (with some of the same authors) suggests that thumb sucking is a potential risk factor, since it is associated more with SIDS infants than with non-SIDS infants (Fleming et al. 1996; L'Hoir et al. 1998a, 1998b). Pacifier sucking reportedly reduces the risk of SIDS by a factor as high as twenty. It would seem, based on this evidence, that pacifier sucking, which is related to breast feeding and can clearly be considered a postnatal experience, reduces the risk of SIDS, whereas thumb sucking, which can be regarded as a prenatal experience, increases the risk of SIDS. These results are quite extraordinary since they are very much in line with the FMD theory. Prenatal, or fetal, activity, like thumb sucking, reminds the infant of being back in the womb, so it should be a risk factor, while postnatal activity, like pacifier sucking, should, according to this theory, act as a preventive measure. At the same time, suggestions that pacifier sucking is preventive because it opens up the oral airway passages cannot explain why thumb sucking has the reverse effect (Cozzi, Cardi, and Cozzi 1998).

Many babies suffer from colic, a condition that keeps them persistently crying. Colic is thought to be caused by spasms in the intestine, often brought on by tension. In an effort to reduce tension in infants, it has been suggested by health officials that one should play a baby tapes of a human heartbeat, or sounds that resemble those it may have experienced in the womb, such as the sound of a washing machine, with fluids swishing back and forth. The idea is that this should comfort the baby by reminding it of being back in the womb, where it was safe and secure. A quick search on the Internet finds many websites that offer tapes and compact discs for sale with these sounds. This is of particular concern to me because it may remind the infant of being back in the womb, which may incite fetal dreams and subsequently lead to SIDS. This may provide another way to test my theory, which would be supported if the incidence of SIDS was higher for groups exposed to these types of sounds over a prolonged period.

There are also a number of cuddly toys on the market that make a sound like a heartbeat. It has been suggested that this may offer a sleeping infant some security in that it might associate the sound with its mother's heartbeat, and possibly with being back in the womb. I regard such toys as dangerous since they may incite fetal dreams. Parents of a newborn baby are usually shown by hospital nursing staff how to wrap their baby so as to make it feel secure and remind it of being back in the womb. If this wrapping does achieve this, I would regard it as potentially dangerous.

I do not support the development of breathing or heart-rate monitors as a means to alert parents if something is wrong. The relationship between conventional sleep apnea and SIDS is unclear, and such devices are usually quite expensive, so they are affordable only to the wealthy and thus largely unavailable to the most vulnerable children from lower socioeconomic groups.

Finally, I would like to stress again that REM sleep seems to serve an important biological function, as exemplified by the facts that almost all mammals have evolved with REM sleep and that lost REM is recovered on subsequent nights of sleep. I would certainly not advocate that infants be deprived of REM sleep in any way, even if it was confirmed that SIDS occurs during REM sleep. REM sleep undoubtedly serves some very important function or functions, especially for a young infant, since it has so much REM sleep. Furthermore, as noted previously, REM sleep may also be important for forgetting potentially harmful fetal memories.

SUMMARY

I have made the specific suggestion that SIDS may be the result of an infant dreaming of being back in the womb, where it did not have to breathe as its mother supplied it with oxygen through the blood. Since dreams are concerned with memory, an infant is likely to dream about its past, when it was in the womb. During dream sleep our bodies try to act out the content of our dreams as much as is possible. My hypothesis is that during REM sleep an infant may recall the breathing pathways that it used as a fetus, stop breathing, and die. This simple hypothesis explains the cause of death, the trigger mechanism, and the simultaneous failure of all arousal mechanisms.

We have seen that the FMD hypothesis is consistent with almost all of the known facts about SIDS, such as the difficulty of determining the cause of death, the occurrence of death during sleep, the risk involved in placing a baby to sleep in the prone position (but not excluding the supine position), the higher incidence during the winter months and in colder climates, and the age-at-death distribution. My theory is consistent with most of the other identified risk factors and the pathological findings.

I have discussed other ways that my theory may be tested, and I have suggested that the best way to reduce the incidence of SIDS is to make the environment of an infant as unlike the womb as possible, so as not to encourage fetal dreams. One can also make changes in the infant's sleeping environment to remind it that it has been born. I hope that others will be able to come up with more ideas that will not only test my theory but also help to reduce the incidence of SIDS.

FURTHER READING

Crib Death: The Sudden Infant Death Syndrome, by Warren Guntheroth. Third edition. Mount Kisco, N.Y.: Futura Publishing Company, 1995. Warren Guntheroth is a respected pediatric cardiologist who has been studying SIDS for nearly forty years. In this book (with some 1,600 references) Dr. Guntheroth reviews the history, literature, pathology, and epidemiology of SIDS, and discusses some of the other theories and ideas relating to SIDS that I have omitted.

References

ACIERNO, R., M. HERSEN, V. VAN HASSELT, G. TREMONT, and K. MEUSER. 1994. "Review of the Validation and Dissemination of Eye-Movement Desensitization and Reprocessing: A Scientific and Ethical Dilemma." *Clinical Psychology Review* 14:287–299.

ADELSON, L., and E. R. KINNEY. 1956. "Sudden and Unexpected Death in Infancy and Childhood." *Pediatrics* 17:663–693.

ALLEN, S. R., W. O. SEILER, H. B. STAHELIN, and R. SPIEGEL. 1987. "Seventy-Two-Hour Polygraphic and Behavioral Recordings of Wakefulness and Sleep in a Hospital Geriatric Unit: Comparison between Demented and Non-Demented Patients." *Sleep* 10:143–159.

ALLISON, T., and D. V. CICCHETTI. 1976. "Sleep in Mammals: Ecological and Constitutional Correlates." *Science* 194:732–734.

ALLISON, T., and H. VAN TWYVER. 1970. "The Evolution of Sleep." *Natural History* 79:56–65.

ALLISON, T., H. VAN TWYVER, and W. R. GOFF. 1972. "Electrophysiological Studies of the Echidna, *Trachyglossus aculeatus*, I: Waking and Sleep." *Archives of Italian Biology* 110:145–184.

AMARI, S. 1972. "Learning Patterns and Pattern Sequences by Self-Organizing Nets of Threshold Elements." *IEEE Transactions on Computers* C21:1197–1206.

AMIT, D. J. 1989. *Modeling Brain Function: The World of Attractor Neural Networks.* Cambridge: Cambridge University Press.

AMIT, D. J. 1995. "The Hebbian Paradigm Reintegrated: Local Reverberations as Internal Representations." *Behavioral and Brain Sciences* 18:617–657.

AMIT, D. J., N. BRUNEL, and M. V. TSODYKS. 1994. "Correlations of Cortical Hebbian Reverberations: Theory versus Experiment." *Journal of Neuroscience* 14:6435–6445.

AMIT, D. J., H. GUTFREUND, and H. SOMPOLINSKY. 1985. "Storing Infinite Number of Patterns in a Spin-Glass Model of Neural Networks." *Physical Review Letters* 55:1530–1535.

ANCH, A. M., C. P. BROWMAN, M. M. MITLER, and J. K. WALSH. 1988. *Sleep: A Scientific Perspective.* Englewood Cliffs, N.J.: Prentice-Hall.

ANDERSON, H. R., and D. G. COOK. 1997. "Passive Smoking and Sudden Infant Death Syndrome—Review of the Epidemiological Evidence." *Thorax* 52:1003–1009.

ASERINSKY, E., and N. KLEITMAN. 1953. "Regularly Occurring Periods of Eye Motility and Concomitant Phenomena during Sleep." *Science* 118:273–274.

ASTON-JONES, G., and F. E. BLOOM. 1981a. "Activity of Norepinephrine-Containing Locus Coeruleus Neurons in Behaving Rats Exhibits Pronounced Response to Non-Noxious Environmental Stimuli." *Journal of Neuroscience* 1:887–900.

ASTON-JONES, G., and F. E. BLOOM. 1981b. "Activity of Norepinephrine-Containing Locus Coeruleus Neurons in Behaving Rats Anticipates Fluctuations in the Sleep-Waking Cycle." *Journal of Neuroscience* 1:876–886.

AUSTRALIAN BUREAU OF STATISTICS. 1990. "Sudden Infant Death Syndrome by States and Territories 1975 to 1988." Canberra: Australian Bureau of Statistics.

BADDELEY, A. 1990. *Human Memory: Theory and Practice.* Needham Heights, Mass.: Allyn and Bacon.

BAGHDOYAN, H. A., A. P. MONACO, M. L. RODRIGO-ANGULO, F. ASSENS, R. W. MCCARLEY, and J. A. HOBSON. 1984. "Microinjection of Neostigmine into the Pontine Reticular Formation of Cats Enhances Desynchronized Sleep Signs." *Journal of Pharmacology and Experimental Therapeutics* 231:173–180.

BEAL, S. M., and R. W. BYARD. 1994. "Accidental Death or Sudden Infant Death Syndrome." *Journal of Paediatrics and Child Health* 30:144–150.

BECKER, L. E. 1990. "Neural Maturation Delay as a Link in the Chain of Events Leading to SIDS." *Canadian Journal of Neurological Sciences* 17:361–371.

BECKWITH, J. B. 1988. "Intrathoracic Petechial Hemorrhages: A Clue to the Mechanism of Death in Sudden Infant Death Syndrome." *Annals of the New York Academy of Sciences* 533:37–47.

BERGER, R. J., P. OLLEY, and I. OSWALD. 1962. "The EEG, Eye Movements and Dreams of the Blind." *Quarterly Journal of Experimental Psychology* 14:183–187.

BERGMAN, A. B. 1970. "Sudden Infant Death Syndrome in King County, Washington: Epidemiologic Aspects." In *Sudden Infant Death Syndrome: Proceedings of the Second International Conference on Causes of Sudden Death in Infants,* ed. A. B. Bergman, J. B. Beckwith, and C. G. Ray, pp. 47–54. Seattle: University of Washington Press.

BLACKMORE, S. 1990. "Dreams That Do What They Are Told." *New Scientist* 125 (1698): 28–31.

BLACKMORE, S. 1999. *The Meme Machine*. Oxford: Oxford University Press.

BLACKMORE, S. 2000. "The Power of Memes." *Scientific American* 283(4): 52–61.

BLAKEMORE, C., and G. F. COOPER. 1970. "Development of the Brain Depends on the Visual Environment." *Nature* 228:477–478.

BLISS, T. V. P., and G. L. COLLINGRIDGE. 1993. "A Synaptic Model of Memory: Long-Term Potentiation in the Hippocampus." *Nature* 361:31.

BLOCH, V., E. HENNEVIN, and P. LECONTE. 1977. "Interaction between Post-Trial Reticular Stimulation and Subsequent Paradoxical Sleep in Memory Consolidation Processes." In *Neurobiology of Sleep and Memory*, ed. R. Drucker-Colin and J. L. McGaugh, pp. 255–272. New York: Academic Press.

BRENNER, R. A., B. G. SIMONSMORTON, B. BHASKAR, N. MEHTA, V. L. MELNICK, M. REVENIS, H. W. BERENDES, and J. D. CLEMENS. 1998. "Prevalence and Predictors of the Prone Sleep Position among Inner-City Infants." *Journal of the American Medical Association* 280:341–346.

BROWMAN, C. P. 1980. "Sleep Following Sustained Exercise." *Psychophysiology* 17: 577–580.

BULTREYS, M. 1990. "High Incidence of Sudden Infant Death Syndrome among Northern Indians and Alaska Natives Compared with Southwestern Indians: Possible Role of Smoking." *Journal of Community Health* 15:185–194.

BYARD, R. W. 1997. "Issues in Diagnosis following the Sudden Infant Death Syndrome." *Journal of Paediatrics and Child Health* 33:467–468.

CAIANIELLO, E. R. 1961. "Outline of a Theory of Thought-Processes and Thinking Machines." *Journal of Theoretical Biology* 11:204–235.

CHANGEUX, J. P. 1986. *Neuronal Man: The Biology of Mind*. Oxford: Oxford University Press. Reprinted 1997. Princeton: Princeton University Press.

CHANGEUX, J. P., P. COURRÈGE, and A. DANCHIN. 1973. "A Theory of the Epigenesis of Neural Networks by Selective Stabilization of Synapses." *Proceedings of the National Academy of Sciences USA* 70:2974–2978.

CHRISTOS, G. A. 1992. "Infant Dreaming: The Fatal Foetal Dream Theory." *Curtin Gazette* 5(4):19–21.

CHRISTOS, G. A. 1993. "Is Alzheimer's Disease Related to a Deficit or Malfunction of Rapid-Eye-Movement REM Sleep?" *Medical Hypotheses* 41:435–439.

CHRISTOS, G. A. 1995a. "Infant Dreaming and Foetal Memory: A Possible Explanation of Sudden Infant Death Syndrome." *Medical Hypotheses* 44:243–250.

CHRISTOS, G. A. 1995b. "On the Origin of Creativity and Its Evolution in the Brain." *Evolution and Cognition* 1:51–53.

CHRISTOS, G. A. 1996. "Investigation of the Crick-Mitchison Reverse-Learning Dream Sleep Hypothesis in a Dynamical Setting." *Neural Networks* 9:427–434.

CHRISTOS, G. A. 1998a. "On the Function of Rapid-Eye-Movement Sleep and Dreaming." *Noetic Journal* 1(2):134–148.

CHRISTOS, G. A. 1998b. "Memory Maintenance in a Model with Alternating Cycles of Learning and Unlearning, As in Awake and Sleep." Curtin University Technical Report 5/98, unpublished.

CHRISTOS, G. A. 1999. "A Possible Basis for Eye-Movement Desensitization (EMD)." *Noetic Journal* 2(2):133–135.

CHRISTOS, G. A., and J. CHRISTOS. 1993. "A Possible Explanation of Sudden Infant Death Syndrome (SIDS)." *Medical Hypotheses* 41:245–246.

CHURCHLAND, P. S. 1986. *Neurophilosophy: Toward a Unified Science of the Mind/Brain.* Cambridge: MIT Press.

CHURCHLAND, P. S., and T. J. SEJNOWSKI. 1992. *The Computational Brain.* Cambridge: MIT Press.

CORBYN, J. A. 1993. *Air Movement in the Human Environment and Sudden Infant Death Syndrome.* Fremantle, Australia: Western Technical Press.

CORNWELL, A. C., P. FEIGENBAUM, and A. KIM. 1998. "SIDS, Abnormal Nighttime REM Sleep and CNS Immaturity." *Neuropediatrics* 29:72–79.

COVELLO, E. 1984. "Lucid Dreaming: A Review and Experiential Study of Waking Intrusions during Stage REM Sleep." *Journal of Mind and Behavior* 5(1):81–98.

COYLE, J. T., D. L. PRICE, and M. R. DELONG. 1983. "Alzheimer's Disease: A Disorder of Cholinergic Innervation." *Science* 219:1184–1190.

COZZI, F., E. CARDI, and D. A. COZZI. 1998. "Dummy Sucking and Sudden Infant Death Syndrome (SIDS)." *European Journal of Pediatrics* 157:952.

CRICK, F., and C. ASANUMA. 1989. "Certain Aspects of the Anatomy and Physiology of the Cerebral Cortex." In *Parallel Distributed Processing: Explorations in the Microstructure of Cognition;* vol. 2, *Psychological and Biological Models,* ed. J. L. McClelland and D. E. Rumelhart, pp. 333–371. Cambridge: MIT Press.

CRICK, F., and G. MITCHISON. 1983. "The Function of Dream Sleep." *Nature* 304:111–114.

CRICK, F., and G. MITCHISON. 1986. "REM Sleep and Neural Net." *Journal of Mind and Behavior* 7:229–249.

CZIKO, G. 1995. *Without Miracles: Universal Selection Theory and the Second Darwinian Revolution.* Cambridge: MIT Press.

DAROFF, R. B., and I. OSORIO. 1984. "The Function of Dreaming." *Neurology* 34:1271.

DAVIES, D. P. 1985. "Cot Death in Hong Kong: A Rare Problem?" *Lancet* 2(8468): 1346–1348.

DAVIES, P. 1983. "Neurotransmitters and Neuropeptides in Alzheimer's Disease." In *Biological Aspects of Alzheimer's Disease,* Banbury report no. 15, ed. R. Katzman. Cold Spring Harbor, N.Y.: Cold Spring Harbor Laboratory.

DAVIES P., and A. J. MALONEY. 1976. "Selective Loss of Cholinergic Neurons in Alzheimer's Disease." *Lancet* 2(8000): 1403.

DAWKINS, R. 1989. *The Selfish Gene,* revised edition. Oxford: Oxford University Press.

DEEG, K. H., U. BETTENDORF, and W. ALDERATH. 1998. "Is Sudden Infant Death a Consequence of Hypofusion of the Brain Stem—First Results of a Doppler Sonographic Study of 23 Infants with Apparent Life Threatening Events." *Monatsschrift Kinderheilkunde* 146:597–602.

DE JONGE, G. A., A. C. ENGLEBERTS, A. J. M. KOOMEN-LIEFTING, and P. J. KOSTENSE. 1989. "Cot Death and Prone Sleeping Position in the Netherlands." *British Medical Journal* 298:722.

DEMENT, W. C. 1960. "The Effect of Dream Deprivation." *Science* 131:1705–1707.

DEMENT, W. C., and N. KLEITMAN. 1957. "The Relation of Eye Movements during Sleep to Dream Activity: An Objective Method for the Study of Dreaming." *Journal of Experimental Psychology* 53:339–346.

DEWAN, E. M. 1968. "The Programming (P) Hypothesis for REM Sleep." *Psychophysiology* 5:365–366.

DITLRICHOVA, J. 1966. "Development of Sleep in Infancy." *Journal of Applied Physiology* 21:1243–1246.

DIX, J. 1998. "Homicide and the Baby-Sitter." *American Journal of Forensic Medicine and Pathology* 19:321–323.

DOUGLAS, A. S., P. J. HELMS, and I. T. JOLLIFFE. 1998. "Seasonality of Sudden Infant Death Syndrome." *Acta Paediatrica* 87:1033–1038.

DWYER, T., and A. L. PONSONBY. 1996. "The Decline in SIDS: A Success Story for Epidemiology." *Epidemiology* 7:323–325.

DWYER, T., A. L. PONSONBY, and D. COUPER. 1999. "Tobacco Smoke Exposure at One Month of Age and Subsequent Risk of SIDS—A Prospective Study." *American Journal of Epidemiology* 149:593–602.

ECCLES, J. C. 1964. *The Physiology of Synapses.* Berlin: Springer-Verlag.

ELLIOT, J., P. VULLERMIN, and P. ROBINSON. 1998. "Maternal Cigarette Smoking Is Associated with Increased Inner Airway Wall Thickness in Children Who Die from Sudden Infant Death Syndrome." *American Journal of Respiratory and Critical Care Medicine* 158:802–806.

ENGLEBERTS, A. C., and G. A. DE JONGE. 1990. "Choice in Sleeping Position for Infants: Possible Association with Cot Death." *Archives of Disease in Childhood* 65:462–467.

FEIGEL'MAN, M. V., and L. B. IOFFE. 1987. "The Augmented Models of Associative Memory: Asymmetric Interaction and Hierarchy of Patterns." *International Journal of Modern Physics* B1:51.

FEWELL, J. E., and F. G. SMITH. 1998. "Perinatal Nicotine Exposure Impairs Ability of Newborn Rats to Auto-Resuscitate from Apnea during Hypoxia." *Journal of Applied Physiology* 85:2066–2074.

FIRSTMAN, R., and J. TALAN. 1997. *The Death of Innocents: A True Story of Murder, Medicine, and the High Stakes of Science*. New York: Bantam Books.

FISHBEIN, W., C. KASTANIOTIS, and D. CHATTMAN. 1974. "Paradoxical Sleep: Prolonged Augmentation following Learning." *Brain Research* 79:61–77.

FLEMING, P. J., P. S. BLAIR, C. BACON, D. BENSLEY, I. SMITH, E. TAYLOR, J. BERRY, J. GOLDING, and J. TRIPP. 1996. "Environment of Infants during Sleep and Risk of the Sudden Infant Death Syndrome: Results of 1993–5 Case Control Study for Confidential Inquiry into Stillbirths and Deaths in Infancy." *British Medical Journal* 313:191–195.

FONTANARI, J. F. 1990. "Generalization in the Hopfield Model." *Journal de Physique Paris* 51:2421–2430.

FONTANARI, J. F., and W. K. THEUMANN. 1990. "On the Storage of Correlated Patterns in Hopfield's Model." *Journal de Physique Paris* 51:375–386.

FORD, R. P. K., E. A. MITCHELL, A. W. STEWART, R. SCRAGG, and B. J. TAYLOR. 1997. "SIDS, Illness, and Acute Medical Care." *Archives of Disease of Childhood* 77:54–55.

FOULKES, D. 1982. *Children's Dreams*. New York: John Wiley and Sons.

FREEMAN, W. J. 1991. "The Physiology of Perception." *Scientific American* 264(2):34–41.

FRENCH, J. W., B. C. MORGAN, and W. G. GUNTHEROTH. 1972. "Infant Monkeys: A Model of Crib Death." *American Journal of Diseases of Children* 123:480–484.

FREUD, S. 1994. *The Interpretation of Dreams*. First published in German in 1900. Transl. A. A. Brill (1911). New York: Alfred A. Knopf.

FRIESS, E., L. TRACHSEL, J. GULDNER, T. SCHIER, A. STEIGER, and F. HOLSBOER. 1995. "DHEA Administration Increases Rapid Eye Movement Sleep and EEG Power in the Sigma Frequency Range." *American Journal of Physiology (Endocrinology and Metabolism)* 268(31):E107–E113.

FUSTER, J. M. 1973. "Unit Activity in Prefrontal Cortex during Delayed-Response Performance: Neuronal Correlates of Transient Memory." *Journal of Neurophysiology* 36:61–78.

GESZTI, T., and F. PAZMANDI. 1989. "Modeling Dream and Sleep." *Physica Scripta* T25:152–155.

GIULIAN, G. G., E. F. GILBERT, and R. L. MOSS. 1987. "Elevated Fetal Hemoglobin Levels in Sudden Infant Death Syndrome." *New England Journal of Medicine* 316:1122–1126.

GOLDBERG, J., R. HORNUNG, T. YAMASHITA, and W. WEHRMACHER. 1986. "Age at Death and Risk Factors in Sudden Infant Death Syndrome." *Australian Paediatric Journal* 22(suppl. 1):21–28.

GOLDING, J. 1997. "Sudden Infant Death Syndrome and Parental Smoking—A Literature Review." *Paediatric and Perinatal Epidemiology* 11:67–77.

GORDON, M. 1987. "Memory Capacity of Neural Networks Learning within Bounds." *Journal de Physique Paris* 48:2053–2058.

GOTO, K., M. MIRMIRAN, M. M. ADAMS, R. V. LONGFORD, R. B. BALDWIN, M. A. BOEDDIKER, and R. L. ARIAGNO. 1999. "More Awakenings and Heart Rate Variability during Supine Sleep in Pre-Term Infants." *Pediatrics* 103:603–609.

GOYCO, P. G., and R. C. BECKERMAN. 1990. "Sudden Infant Death Syndrome." *Current Problems in Pediatrics* 20:297–346.

GREEN, J. D., and A. A. ARDUINI. 1954. "Hippocampal Electrical Activity in Arousal." *Journal of Neurophysiology* 17:533–557.

GRETHER, J. K., and J. SCHULMAN. 1989. "Sudden Infant Death Syndrome and Birth Weight." *Journal of Pediatrics* 114:561–567.

GRINIASTY, M., M. V. TSODYKS, and D. J. AMIT. 1993. "Conversion of Temporal Correlations between Stimuli to Spatial Correlations between Attractors." *Neural Computation* 5:1–17.

GUILLEMINAULT, C., R. PERAITA, M. SOUQUET, and W. C. DEMENT. 1975. "Apneas during Sleep in Infants: Possible Relationship with SIDS." *Science* 190:677–679.

GUNTHEROTH, W. G. 1995. *Crib Death: The Sudden Infant Death Syndrome*. 3rd Edition. Mount Kisco, N.Y.: Futura Publishing Company.

GUNTHEROTH, W. G., and I. KAWABORI. 1975. "Hypoxic Apnea and Gasping." *Journal of Clinical Investigation* 56:1371–1377.

GUNTHEROTH, W. G., and P. S. SPIERS. 2001. "Thermal Stress in Sudden Infant Death: Is There Ambiguity with Rebreathing Hypothesis." *Pediatrics* 107:693–698.

GUTFREUND, H. 1990. "The Effect of Synaptic Asymmetry in Attractor Neural Networks." In *Neural Networks and Spin Glasses,* ed. W. K. Theumann and R. Koberle, pp. 49–66. Singapore: World Scientific.

HAGAN, L. L., D. W. GOETZ, C. H. REVERCOMB, and J. GARRIOTT. 1998. "Sudden Infant Death Syndrome—A Search for Allergen Hypersensitivity." *Annals of Allergy, Asthma, & Immunology* 80:227–231.

HATA, K., M. FUNAYAMA, S. TOKUDOME, and M. MORITA. 1997. "Problems on the Diagnosis of Sudden Infant Death Syndrome." *Acta Paediatrica Japonica* 39:559–565.

HEBB, D. A. 1949. *The Organization of Behavior.* New York: Wiley and Sons.

HENDERSON-SMART, D., A. L. PONSONBY, and E. MURPHY. 1998. "Reducing the Risk of Sudden Infant Death Syndrome—A Review of the Scientific Literature." *Journal of Paediatrics and Child Health* 34:213–219.

HERTZ, J. A., G. GRINSTEIN, and S. A. SOLLA. 1987. "Irreversible Spin Glasses and Neural Networks." In *Heidelberg Colloquium on Glassy Dynamics,* ed. J. L. van Hemmen and I. Morgenstern, pp. 538–546. Heidelberg: Springer-Verlag.

HILLMAN, L. S. 1991. "Theories and Research." In *Sudden Infant Death Syndrome,* ed. C. A. Corr, H. Fuller, C. A. Barnickol, and D. M. Corr, pp. 14–41. New York: Springer Publishing Company.

HOBSON, J. A. 1988. *The Dreaming Brain.* New York: Basic Books.

HOBSON, J. A. 1989. *Sleep.* Scientific American Library. New York: W. H. Freeman and Company.

HOBSON, J. A. 1990. "Sleep and Dreaming." *Journal of Neuroscience* 10:371–382.

HOBSON, J. A. 1998. *Consciousness.* Scientific American Library. New York: W. H. Freeman and Company.

HOBSON, J. A., and R. W. MCCARLEY. 1977. "The Brain as a Dream State Generator: An Activation-Synthesis Hypothesis of the Dream Process." *American Journal of Psychiatry* 134:1335–1348.

HOBSON, J. A., R. W. MCCARLEY, and P. W. WYZINSKI. 1975. "Sleep Cycle Oscillations: Reciprocal Discharge by Two Brainstem Neuronal Groups." *Science* 189:55–58.

HOFFMAN, H., K. DAMUS, L. HILLMAN, and E. KRONGRAD. 1988. "Risk Factors for SIDS: Results of the National Institute of Child Health and Human Development SIDS Cooperative Epidemiological Study." *Annals of the New York Academy of Sciences* 533:13–30.

HOOPER, J., and D. TERESI. 1992. *The 3-Pound Universe.* New York: Jeremy P. Tarcher, Perigee Books.

HOPFIELD, J. J. 1982a. "Neural Networks and Physical Systems with Emergent Collective Computational Abilities." *Proceedings of the National Academy of Sciences USA* 79:2554–2558.

HOPFIELD, J. J. 1982b. "Brain, Computer and Memory." *Engineering and Science* September, 2–7.

HOPFIELD, J. J. 1984a. "Collective Processing and Neural States." In *Modeling and Analysis in Biomedicine*, ed. C. Nicollini. pp. 369–389. New York: World Scientific.

HOPFIELD, J. J. 1984b. "Neurons with Graded Response Have Collective Computational Properties like Those of Two-State Neurons." *Proceedings of the National Academy of Sciences USA* 81:3088–3092.

HOPFIELD, J. J. 1995. "Pattern Recognition Computation Using Action Potential Timing for Stimulus Representation." *Nature* 376:33–36.

HOPFIELD, J. J., and C. BRODY. 2000. "What Is a Moment? 'Cortical' Sensory Integration over a Brief Interval." *Proceedings of the National Academy of Sciences USA* 97:13919–13924.

HOPFIELD, J. J., and C. BRODY. 2001. "What is a moment? Transcient Synchrony as a Collective Mechanism for Spatiotemporal Integration." *Proceedings of the National Academy of Sciences USA* 98:1282–1287.

HOPFIELD, J. J., D. I. FEINSTEIN, and R. G. PALMER. 1983. "Unlearning Has a Stabilising Effect in Collective Memories." *Nature* 304:158–159.

HOPPERBROUWERS, T., J. HODGMAN, K. ARAKAWA, S. A. GEIDEL, and M. B. STERMAN. 1988. "Sleep and Waking States in Infancy: Normative Studies." *Sleep* 11:387–401.

HORNE, J. A. 1988. *Why We Sleep: The Function of Sleep in Humans and Other Mammals.* Oxford: Oxford University Press.

HORNE, J. A., and M. J. MCGRATH. 1984. "The Consolidation Hypothesis for REM Sleep Function: Stress and Other Confounding Factors—A Review." *Biological Psychology* 18:165–184.

HUBEL, D. H. 1988. *Eye, Brain, and Vision.* Scientific American Library. New York: W. H. Freeman and Company.

HUBEL, D. H., and T. N. WIESEL. 1970. "The Period of Susceptibility to the Physiological Effects of Unilateral Eye Closure in Kittens." *Journal of Physiology* 206:419–436.

HUNT, L., P. FLEMING, and J. GOLDING. 1997. "Does the Supine Sleeping Position Have Any Adverse Effects on the Child? I: Health in the First Six Months." *Pediatrics* 100:111–119.

JACOBS, B. L., and M. E. TRULSON. 1979. "Dreams, Hallucinations and Psychosis—The Serotonin Connection." *Trends in Neuroscience* 2:276–280.

JAMES, W. 1890. *The Principles of Psychology.* Reprinted in 1950. New York: Dover.

JORM, A. F., B. RODGERS, P. A. JACOMB, H. CHRISTENSEN, S. HENDERSON, and A. E. KORTEN. 1999. "Smoking and Mental Health: Results from a Community Survey." *Medical Journal of Australia* 170:74–77.

JUNG, C. G. 1976. *Modern Man in Search of a Soul.* First published in German in 1933. Transl. C. F. Baynes and W. S. Dell. New York: Harcourt, Brace and Company.

KALES, A., F. S. HOEDEMAKER, A. JACOBSON, and E. L. LICHTENSTEIN. 1964. "Dream Deprivation: An Experimental Reappraisal." *Nature* 204:1337–1338.

KALES, A., and J. D. KALES. 1974. "Sleep Disorders: Recent Findings in the Diagnosis and Treatment of Disturbed Sleep." *New England Journal of Medicine* 290:487–499.

KANDEL, E. R., J. H. SCHWARTZ, and T. M. JESSELL. (Eds.) 1991. *Principles of Neural Science.* Englewood Cliffs, N.J.: Prentice-Hall.

KARNI, A., D. TANNE, B. S. RUBINSTEIN, J. J. M. ASKENASI, and D. SAGI. 1994. "Dependence on REM Sleep of Overnight Improvement of a Perceptual Skill." *Science.* 265:679–682.

KATZ, B. 1969. *The Release of Neural Transmitter Substances.* Liverpool: Liverpool University Press.

KINNEY, H. C., B. A. BRODY, D. M. FINKELSTEIN, G. F. VAWTER, F. MANDELL, and F. H. GILLES. 1991. "Delayed Central Nervous System Myelination in the Sudden Infant Death Syndrome." *Journal of Neuropathology and Experimental Neurology* 50:29–48.

KINNEY, H. C., and J. J. FILIANO. 1988. "Brainstem Research in the Sudden Infant Death Syndrome." *Pediatrician* 15:240–250.

KINNEY, H. C., J. J. FILIANO, S. F. ASSMANN, F. MANDELL, M. VALDES-DAPENA, H. F. KROUS, T. ODONNELL, L. A. RAVA, and W. F. WHITE. 1998. "Tritiated-Naloxone Binding to Brainstem Opioid Receptors in the Sudden Infant Death Syndrome." *Journal of the Autonomic Nervous System* 69:156–163.

KLEINFELD, D. 1986. "Sequential State Generation by Model Neural Networks." *Proceedings of the National Academy of Sciences USA* 83:9469–9473.

KLEINFELD, D., and D. B. PENDERGRAFT. 1987. "Unlearning Increases the Storage Capacity of Content Addressable Memories." *Biophysicial Journal* 51:47–53.

KLEITMAN, N. 1963. *Sleep and Wakefulness.* Chicago: University of Chicago Press.

KOHLENDORFER, U., E. HABERLANDT, S. KIECHL, and W. SPERL. 1997. "Pre- and Postnatal Medical Care and Risk of Sudden Infant Death Syndrome." *Acta Paediatrica* 86:600–603.

KROUS, H. F. 1988. "Pathological Considerations of Sudden Infant Death Syndrome." *Pediatrician* 15:231–239.

LaBerge, S. 1981. "Lucid Dreaming: Directing the Action as it Happens." *Psychology Today* 15(1): 48–57.

LaBerge, S. 1986. *Lucid Dreaming: The Power of Being Awake and Aware in Your Dreams.* New York: Ballantine Books.

LaBerge, S. P., and W. C. Dement. 1982a. "Voluntary Control of Respiration during REM Sleep." *Sleep Research* 11:107.

LaBerge, S. P., and W. C. Dement. 1982b. "Lateralization of Alpha Activity for Dreamed Singing and Counting During REM Sleep." *Psychophysiology* 19: 331–332.

Landauer, T. K. 1986. "How Much Do People Remember? Some Estimates of the Quantity of Learned Information in Long-Term Memory." *Cognitive Science* 10:477–493.

Lavie, P., H. Pratt, B. Scharf, R. Peled, and J. Brown. 1984. "Localized Pontine Lesion: Nearly Total Absence of REM Sleep." *Neurology* 34:118–120.

Lee, N. N. Y., Y. T. Chan, D. P. Davies, E. Lau, and D. C. P. Yip. 1989. "Sudden Infant Death Syndrome in Hong Kong: Confirmation of Low Incidence." *British Medical Journal* 298:721.

Ledwidge, M., G. Fox, and T. Mathews. 1998. "Neurocardiogenic Syncope—A Model for SIDS." *Archives of Disease in Childhood* 78:481–483.

L'Hoir, M. P., A. C. Engelberts, G. T. J. van Well, T. Bajanowski, K. Helweg-Larsen, and J. Huber. 1998a. "Sudden Unexpected Death in Infancy: Epidemiologically Determined Risk Factors Related to Pathological Classification." *Acta Paediatrica* 87:1279–1287.

L'Hoir, M. P., A. C. Engleberts, G. T. J. van Well, S. McClelland, P. Westers, T. Dandachli, G. J. Mellenbergh, W. H. G. Wolters, and J. Huber. 1998b. "Risk and Preventative Factors for Cot Death in the Netherlands, a Low-Incidence Country." *European Journal of Pediatrics* 157:681–688.

L'Hoir, M. P., A. C. Engleberts, G. T. J. van Well, P. Westers, G. J. Mellenbergh, W. H. G. Wolters, and J. Hubers. 1998c. "Case-Control Study of Current Validity of Previously Described Risk Factors for SIDS in the Netherlands." *Archives of Disease in Childhood* 79:386–393.

Lindgren, C., J. M. D. Thompson, L. Haggblom, and J. Milerad. 1998. "Sleeping Position, Breastfeeding, Bedsharing and Passive Smoking in 3-Month-Old Swedish Infants." *Acta Paediatrica* 87:1028–1032.

Little, G. A., R. B. Ariagno, B. Beckwith, J. G. Brooks, W. G. Guntheroth, J. E. Hodgman, J. Kattwinkel, D. H. Kelly, F. Mandell, B. L. McEntire, T. A. Merritt, N. M. Nelson, D. W. Nielson, D. C. Shannon, A. R. Spitzer, A. Steinschneider, M. Valdes-Dapena, and S. Weinstein. 1985. "Prolonged

Infantile Apnea: 1985." American Academy of Pediatrics. Task Force on Prolonged Infantile Apnea. *Pediatrics* 76:129–131.

LITTLE, W. A. 1974. "The Existence of Persistent States in the Brain." *Mathematical Biosciences* 19:101–120.

LITTLE, W. A., and G. W. SHAW. 1978. "Analytic Study of the Memory Storage Capacity of a Neural Network." *Mathematical Biosciences* 39:281–290.

LIVANOVA, A. 1980. *Landau, A Great Physicist and Teacher.* New York: Pergamon Press.

LOBBAN, C. D. R. 1991. "The Human Dive Reflex as a Primary Cause of SIDS." *Medical Journal of Australia* 155:561–563.

MACDORMAN, M. F., S. CNATTINGIUS, H. J. HOFFMAN, M. S. KRAMER, and B. HAGLUND. 1997. "Sudden Infant Death Syndrome and Smoking in the United States and Sweden." *American Journal of Epidemiology* 146:249–257.

MALLOY, M. H., J. C. KLEINMAN, G. H. LAND, and W. F. SCHRAMM. 1988. "The Association of Maternal Smoking with Age and Cause of Infant Death." *American Journal of Epidemiology* 128:46–55.

MARCHAL, C., and P. DROULLE. 1988. "Fetal Respiratory Movements." *Revue des Maladies Respiratoires* 5:207–212.

MARQUIS, J. N. 1991. "A Report on Seventy-Eight Cases Treated by Eye Movement Desensitization." *Journal of Behavior Therapy and Experimental Psychiatry* 22:187–192.

MAURICE, D. M. 1998. "The Von Sallmann Lecture 1996: An Ophthalmological Explanation of REM Sleep." *Experimental Eye Research* 66:139–145.

MCGRATH, M. J., and D. B. COHEN. 1978. "REM Sleep Facilitation of Adaptive Waking Behavior: A Review of the Literature." *Psychology Bulletin* 85:24–57.

MEYER, J. S., L. A. HAYMAN, T. AMANO, S. NAKAJIMA, T. SHAW, P. LARZON, S. DERMAN, I. KARACAN, and Y. HARATI. 1981. "Mapping Local Blood Flow of Human Brain by CT Scanning during Stable Xenon Inhalation." *Stroke* 12:426–436.

MEZARD, M., J. P. NADAL, and G. TOULOUSE. 1986. "Solvable Models of Working Memories." *Journal de Physique Paris* 47:1457–1462.

MILLER, G. A. 1956. "The Magical Number Seven Plus or Minus Two: Some Limits on Our Capacity for Processing Information." *Psychological Review* 63:81–97.

MITCHELL, E. A., M. CLEMENTS, S. M. WILLIAMS, A. W. STEWART, A. CHENG, and R. P. K. FORD. 1999. "Seasonal Differences in Risk Factors for Sudden Infant Death Syndrome." *Acta Paediatrica* 88:253–258.

MITCHELL, E. A., L. SCRAGG, and M. CLEMENTS. 1996. "Soft Cot Mattresses and the Sudden Infant Death Syndrome." *New Zealand Medical Journal* 109:206–207.

MITCHELL, E. A., R. SCRAGG, A. W. STEWART, D. M. O. BECROFT, B. J. TAYLOR, R. P. K. FORD, I. B. HASSALL, D. M. J. BARRY, E. M. ALLEN, and A. P. ROBERTS. 1991. "Results from the First Year of the New Zealand Cot Death Study." *New Zealand Medical Journal* 104:71–76.

MITCHELL, E. A., and A. W. STEWART. 1997. "Gender and Sudden Infant Death Syndrome." *Acta Paediatrica* 86:854–856.

MITCHELL, E. A., J. M. D. THOMPSON, R. P. K. FORD, and B. J. TAYLOR. 1998. "Sheepskin Bedding and the Sudden Infant Death Syndrome." *Journal of Pediatrics* 133:701–704.

MIYASHITA, Y. 1988. "Neuronal Correlate of Visual Associative Long-Term Memory in the Primate Temporal Cortex." *Nature* 335:817–820.

MIYASHITA, Y., and H. S. CHANG. 1988. "Neuronal Correlate of Pictorial Short-Term Memory in the Primate Temporal Cortex." *Nature* 331:68–70.

MORRISON, A. R. 1983. "A Window on the Sleeping Brain." *Scientific American* 248(4):86–94.

MOSKO, S., C. RICHARD, and J. MCKENNA. 1997. "Infant Arousals during Mother-Infant Bed Sharing—Implications for Infant Sleep and Sudden Infant Death Syndrome." *Pediatrics* 100:841–849.

MOUNTCASTLE, V. B. 1957. "Modality and Topographic Properties of Single Neurons of Cat's Somatic Sensory Cortex." *Journal of Neurophysiology* 20:408–434.

MUKHAMETOV, L. M. 1984. "Sleep in Marine Mammals." In *Sleep Mechanisms,* ed. A. A. Borbely, and J. L. Valatx, pp. 227–238. Berlin: Springer-Verlag.

NACHMANOFF, D. B., A. PANIGRAHY, J. J. FILIANO, F. MANDELL, L. A. SLEEPER, M. VALDES-DAPENA, H. F. KROUS, W. F. WHITE, and H. C. KINNEY. 1998. "Brainstem H-3-Nicotine Binding in the Sudden Infant Death Syndrome." *Journal of Neuropathology and Experimental Neurology* 57:1018–1035.

NADEL, J. P., G. TOULOUSE, J. P. CHANGEUX, and S. DEHAENE. 1986. "Networks of Formal Neurons and Memory Palimpsests." *Europhysics Letters* 1:535–542.

NAEYE, R. L. 1980. "Sudden Infant Death." *Scientific American* 242:52–56.

NAUTA, W. J. H., and M. FEIRTAG. 1979. "The Organization of the Brain." In *The Brain,* pp. 40–55. A Scientific American Book. San Francisco: W. H. Freeman and Company.

NELSON, E. A., and B. J. TAYLOR. 1988. "Climatic and Social Associations with Post-neonatal Mortality in New Zealand." *New Zealand Medical Journal* 101:443–446.

NELSON, E. A. S., B. J. TAYLOR, and I. L. WEATHERALL. 1989. "Sleeping Position and Infant Bedding May Predispose to Hyperthermia and Sudden Infant Death Syndrome." *Lancet* 1(8631):199–201.

NICOL, S. C., N. A. ANDERSEN, N. H. PHILLIPS, and R. J. BERGER. 2000. "The Echidna Manifests Typical Characteristics of Rapid Eye Movement Sleep." *Neuroscience Letters* 283:49–52.

OBONAI, T., M. YASUHARA, T. NAKAMURA, and S. TAKASHIMA. 1998. "Catecholamine Neurons Alteration in the Brainstem of Sudden Infant Death Syndrome Victims." *Pediatrics* 101:285–288.

ORIOT, D., M. BERTHIER, J. P. SAULNIER, D. BLAY, J. P. FOHR, V. VUILLERME, and J. B. SAULNIER. 1998. "Prone Position May Increase Temperature around the Head of the Infant." *Acta Paediatrica* 87:1005–1007.

OSORIO, I., and R. DAROFF. 1980. "Absence of REM and Altered NREM Sleep in Patients with Spinocerebellar Degeneration and Slow Saccades." *Annals of Neurology* 7:277–280.

OYEN, N., T. MARKESTAD, R. SKJAERVEN, L. M. IRGENS, K. HELWEGLARSEN, B. ALM, G. NORVENIUS, and G. WENNERGREN. 1997. "Combined Effects of Sleeping Position and Prenatal Risk Factors in Sudden Infant Death Syndrome—The Nordic Epidemiological SIDS Study." *Pediatrics* 100:613–621.

PAMPHLETT, R., J. RAISANEN, and S. KUM-JEW. 1999. "Vertebral Artery Compression Resulting from Head Movement: A Possible Cause of Sudden Infant Death Syndrome." *Pediatrics* 103:460–468.

PANIGRAHY, A., J. J. FILIANO, L. A. SLEEPER, F. MANDELL, M. VALDES-DAPENA, H. F. KROUS, L. A. RAVA, W. F. WHITE, and H. C. KINNEY. 1997. "Decreased Kainate Receptor Binding in the Arcuate Nucleus of the Sudden Infant Death Syndrome." *Journal of Neuropathology and Experimental Neurology* 56:1253–1261.

PARISI, G. 1986a. "A Memory Which Forgets." *Journal of Physics A: Mathematical and General* 19:L617–L620.

PARISI, G. 1986b. "Asymmetric Neural Networks and the Process of Learning." *Journal of Physics A: Mathematical and General* 19:L675–L680.

PARMELEE, A. H., W. H. WENNER, Y. AKIYAMA, M. SCHULTZ, and E. STERN. 1967. "Sleep States in Premature Infants." *Developmental Medicine and Child Neurology* 9:70–77.

PARMELEE, A. H., W. H. WENNER, and H. R. SCHULTZ. 1964. "Infant Sleep Patterns: From Birth to 16 Weeks of Age." *Journal of Pediatrics* 65:576–582.

PERETTO, P. 1988. "On Models of Short and Long Term Memories." In *Computer Simulation in Brain Science*, ed. R. Cottrill, pp. 88–103. Cambridge: Cambridge University Press.

PERETTO, P. 1992. *An Introduction to the Modeling of Neural Networks.* Cambridge: Cambridge University Press.

PERRY, G. W., R. VARGASCUBA, and R. P. VERTES. 1997. "Fetal Hemoglobin Levels in Sudden Infant Death Syndrome." *Archives of Pathology and Laboratory Medicine* 121:1048–1054.

PETERSON, D. R. 1988. "Clinical Implications of Sudden Infant Death Syndrome Epidemiology." *Pediatrician* 15:198–203.

PETERSON, D. R., N. M. CHIN, and L. D. FISHER. 1980. "The Sudden Infant Death Syndrome: Repetitions in Families." *Journal of Pediatrics* 97:265–267.

PETERSON, D. R., E. E. SABOTTA, and D. STRICKLAND. 1988. "Sudden Infant Death Syndrome in Epidemiological Perspective: Etiologic Implications of Variation with Season of the Year." In *Cardiac and Respiratory Mechanisms and Interventions,* ed. P. J. Schwartz, D. P. Southall, and M. Valdes-Dapena, pp. 6–12. *Annals of the New York Academy of Sciences,* vol. 533.

POETS, C. F., R. G. MENY, M. R. CHOBANIAN, and R. E. BONOFIGLO. 1999. "Gasping and Other Cardiorespiratory Patterns during Sudden Infant Deaths." *Pediatric Research* 45:350–354.

POINCARÉ, H. 1982. *The Foundations of Science.* First published in French in 1908. Transl. G. B. Halstead. Washington, D.C.: University Press of America.

POMPEIANO, O. 1979. "Cholinergic Activation of Reticular and Vestibular Mechanisms Controlling Posture and Eye Movements." In *The Reticular Formation Revisited,* ed. J. A. Hobson and M. A. B. Braxier, pp. 473–572. New York: Raven.

PONSONBY, A. L., T. DWYER, D. COUPER, and J. COCHRANE. 1998. "Association between Use of a Quilt and Sudden Infant Death Syndrome." *British Medical Journal* 316:195–196.

PONSONBY, A. L., T. DWYER, S. V. KASL, and J. A. COCHRAINE. 1995. "The Tasmanian SIDS Case-Control Study: Univariable and Multivariable Risk Factor Analysis." *Paediatric Perinatal Epidemiology* 9:256–272.

PONSONBY, A. L., M. E. JONES, J. LUMLEY, T. DYWER, and N. GILBERT. 1992. "Climatic Temperature and Variation in the Incidence of Sudden Infant Death Syndrome between the Australian States," *Medical Journal of Australia* 156:246–248.

RECHTSCHAFFEN, A., R. WATSON, M. Z. WINCOR, S. MOLINARI, and S. G. BARTA. 1972. "The Relationship of Phasic and Tonic Periorbital EMG Activity to NREM Mentation." *Sleep Research* 1:114.

ROBINS, A., and S. MCCALLUM. 1999. "The Consolidation of Learning during Sleep: Comparing the Pseudorehearsal and Unlearning Accounts." *Neural Networks* 12:1191–1206.

ROCHESTER, N., J. H. HOLLAND, L. H. HAIBT, and W. L. DUDA. 1956. "Tests on a Cell Assembly Theory of the Action of the Brain, Using a Large Digital Computer." *IRE Transactions on Information Theory* IT-2:80–93.

ROFFWARG, H. P., J. H. HERMAN, C. BOWE-ANDERS, and E. S. TAUBER. 1978. "The Effects of Sustained Alterations of Waking Visual Input on Dream Content." In *The Mind in Sleep*, ed. A. M. Arkis, J. S. Antrobus, and S. J. Ellman, pp. 295–349. Hillsdale, N.J.: Lawrence Erlbaum Associates.

ROFFWARG, H. P., J. N. MUZIO, and W. C. DEMENT. 1966. "Ontogenetic Development of the Human Sleep-Dream Cycle." *Science* 152:604–619.

SAKAI, K., and Y. MIYASHITA. 1991. "Neural Organization for the Long-Term Memory of Paired Associates." *Nature* 354:152–155.

SAWCZENKO, A., and P. J. FLEMING. 1996. "Thermal Stress, Sleeping Position, and Sudden Infant Death Syndrome." *Sleep* 19 (suppl.):S267–S270.

SCHECHTMAN, V. L., R. M. HARPER, K. A. KLUGE, A. J. WILSON, H. J. HOFFMAN, and D. P. SOUTHALL. 1988. "Cardiac and Respiratory Patterns in Normal Infants and Victims of Sudden Infant Death Syndrome." *Sleep* 11:413–424.

SCHECHTMAN, V. L., R. M. HARPER, A. J. WILSON, and D. P. SOUTHALL. 1992. "Sleep State Organization in Normal Infants and Victims of Sudden Infant Death Syndrome." *Pediatrics* 89:865–870.

SCHECHTMAN, V. L., M. Y. LEE, A. J. WILSON, and R. M. HARPER. 1996. "Dynamics of Respiratory Patterning in Normal Infants and Infants Who Subsequently Died of Sudden Infant Death Syndrome." *Pediatric Research* 40:571–577.

SCHEERS, N. J., C. M. DAYTON, and J. S. KEMP. 1998. "Sudden Infant Death with External Airways Covered—Case Comparison Study of 206 Deaths in the United States." *Archives of Pediatrics & Adolescent Medicine* 152:540–547.

SCHLUTER, P. J., R. P. K. FORD, E. A. MITCHELL, and B. J. TAYLOR. 1998. "Residential Mobility and Sudden Infant Death Syndrome." *Journal of Paediatrics and Child Health* 34:432–437.

SCHWARTZ, P. J., M. STRAMBABADIALE, A. SEGANTINI, P. AUSTONI, G. BOSI, R. GIORGETTI, F. GRANCINI, E. D. MARNI, F. PERTICONE, D. ROSTIS, and P. SALICE. 1998. "Prolongation of the QT Interval and Sudden Infant Death Syndrome." *New England Journal of Medicine* 338:1709–1714.

SCOVILLE, W. B., and B. MILNER. 1957. "Loss of Recent Memory after Bilateral Hippocampus Lesions." *Journal of Neurology, Neurosurgery and Psychiatry* 20:11-21.

SCRAGG, R. K. R., and E. A. MITCHELL. 1998. "Side Sleeping and Bed Sharing in the Sudden Infant Death Syndrome." *Annals of Medicine* 30:345–349.

SCRAGG, R. K. R, E. A. MITCHELL, A. W. STEWART, R. P. K. FORD, B. J. TAYLOR, I. B. HASSALL, S. M. WILLIAMS, and J. M. D. THOMPSON. 1996. "Infant Room-Sharing and Prone Sleep Position in Sudden Infant Death Syndrome." *Lancet* 347:7–12.

SHAPIRO, C. M., R. BORTZ, D. MITCHELL, P. BARTEL, and P. JOOSTE. 1981. "Slow-Wave Sleep: A Recovery Period after Exercise." *Science* 214:1253–1254.

SHAPIRO, F. 1989. "Eye Movement Desensitization: A New Treatment for Post-Traumatic Stress Disorder." *Journal of Behavior Therapy and Experimental Psychiatry* 20:211–217.

SHRIVASTAVA, A., P. DAVIS, and D. P. DAVIES. 1997. "SIDS—Parental Awareness and Infant Care Practices in Contrasting Socioeconomic Classes in Cardiff." *Archives of Disease in Childhood* 77:52–53.

SLOTKIN, T. A. 1998. "Fetal Nicotine or Cocaine Exposure—Which One is Worse?" *Journal of Pharmacology and Experimental Therapeutics* 285:931–945.

SMITH, C. 1985. "Sleep States and Learning: A Review of the Animal Literature." *Neuroscience and Biobehavioral Reviews* 9:157–168.

SMITH, C., K. KITAHAMA, J. L. VALATX, and M. JOUVET. 1974. "Increased Paradoxical Sleep in Mice during Acquisition of a Shock Avoidance Task." *Brain Research* 77:221–230.

SMITH, C., and L. LAPP. 1991. "Increases in Number of REMs and REM Density in Humans following an Intensive Learning Period." *Sleep* 14:325–330.

SMITH, G. E. 1902. "Mammalia, Order Monotremata." In *Catalogue of the Physiological Series of Comparative Anatomy*, vol. 2, pp. 138–157. London: Museum of the Royal College of Surgeons.

SNYDER, S. 1986. *Drugs and the Brain*. Scientific American Library. New York: W. H. Freeman and Company.

SOMPOLINSKY, H., and I. KANTER. 1986. "Temporal Association in Asymmetric Neural Networks." *Physical Review Letters* 57:2861–2864.

SOUTHALL, D. P. 1999. "Examine Data in Schwartz Article with Extreme Care." *Pediatrics* 103:819–820.

SOUTHALL, D. P., W. A ARROWSMITH, V. STEBBENS, and J. R. ALEXANDER. 1986. "QT Interval Measurements before Sudden Infant Death Syndrome." *Archives of Disease in Childhood* 61:327–333.

SOUTHALL, D. P., J. RICHARDS, D. J. BROWN, P. G. B. JOHNSTON, M. DESWIET, and E. A. SHINEBOURNE. 1980. "24-Hour Tape Recordings of ECG and Respiration in the Newborn Infant with Findings Related to Sudden Death and Unexplained Brain Damage." *Archives of Disease in Childhood* 55:7–16.

SOUTHALL, D. P., J. M. RICHARDS, K. J. RHODEN, J. R. ALEXANDER, E. A. SHINEBOURNE, W. A. ARROWSMITH, J. E. CREE, P. J. FLEMING, A. GONCALVES, and

R. L'E. ORME. 1982. "Prolonged Apnea and Cardiac Arrhythmias in Infants Discharged from Neonatal Intensive Care Units: Failure to Predict an Increased Risk for Sudden Infant Death Syndrome." *Pediatrics* 70:844–851.

SPERLING, G. 1960. "The Information Available in Brief Visual Presentations." *Psychological Monographs* 27:285–292.

SPIERS, P. S., and W. G. GUNTHEROTH. 1997. "The Seasonal Distribution of Infant Deaths by Age—A Comparison of Sudden Infant Death Syndrome and Other Causes of Death." *Journal of Paediatrics and Child Health* 33:408–412.

SQUIRE, L. R., and E. R. KANDEL. 1999. *Memory: From Mind to Molecules.* Scientific American Library. New York: W. H. Freeman and Company.

STEELE, R. 1970. "Sudden Infant Death Syndrome in Ontario, Canada: Epidemiologic Aspects." In *Sudden Infant Death Syndrome: Proceedings of the Second International Conference on Causes of Sudden Death in Infants,* ed. A. B. Bergman, J. B. Beckwith, and C. G. Ray, pp. 64–72. Seattle: University of Washington Press.

STEINSCHNEIDER, A. 1972. "Prolonged Apnea and Sudden Infant Death Syndrome: Clinical and Laboratory Observations." *Pediatrics* 50:646–654.

STORM, H., G. NYLANDER, and O. D. SAUGSTAD. 1999. "The Amount of Brainstem Gliosis in Sudden Infant Death Syndrome SIDS Victims Correlates with Maternal Cigarette Smoking during Pregnancy." *Acta Paediatrica* 88:13–18.

SUNDERLAND, T., C. R. MERRIL, M. G. HARRINGTON, B. A. LAWLOR, S. E. MOLCHAN, R. MARTINEZ and D. L. MURPHY. 1989. "Reduced Plasma Dehydroepiandrosterone Concentrations in Alzheimer's Disease." *Lancet* 2(8662):570.

TAYLOR, B. J., S. M. WILLIAMS, E. A. MITCHELL, R. P. K. FORD, D. M. O. BECROFT, A. W. STEWART, R. SCRAGG, I. B. HASSALL, E. M. ALLEN, D. M. J. BARRY, and A. P. ROBERTS. 1996. "Symptoms, Sweating and Reactivity of Infants Who Die of SIDS Compared with Community Controls." *Journal of Paediatrics and Child Health* 32:316–322.

TOULOUSE, G., S. DEHAENE, and J. P. CHANGEUX. 1986. "Spin Glass Model of Learning by Selection." *Proceedings of the National Academy of Sciences USA* 83:1695–1698.

TSODYKS, M. V. 1989. "Associative Memory in Neural Networks with the Hebbian Learning Rule." *Modern Physics Letters B* 37:555–560.

VALDES-DAPENA, M. A. 1983. "The Morphology of the Sudden Infant Death Syndrome: An Overview." In *Sudden Infant Death Syndrome,* ed. J. T. Tildon, L. M. Roeder, and A. Steinschneider, pp. 169–182. New York: Academic Press.

VALDES-DAPENA, M. A. 1988. "Sudden Infant Death Syndrome: Overview of Recent Research Developments from a Pediatric Pathologist's Perspective." *Pediatrician* 5:222–230.

VALDES-DAPENA, M. A. 1991. "The Phenomenon of Sudden Infant Death Syndrome and Its Challenges." In *Sudden Infant Death Syndrome,* ed. C. A. Corr, H. Fuller, C. A. Barnickol, and D. M. Corr, pp. 2–13. New York: Springer.

VAN HEMMEN, J. L. 1997. "Hebbian Learning, Its Correlation Catastrophe, and Unlearning." *Network: Computation in Neural Systems* 8:V1–V17.

VAN HEMMEN, J. L., L. B. IOFFE, R. KUHN, and M. VASS. 1990. "Increasing the Efficiency of a Neural Network through Unlearning." *Physica A* 163:386–392.

VOGEL, G. W. 1968. "REM Deprivation III: Dreaming and Psychosis." *Archives of General Psychiatry* 18:312–329.

VOGEL, G. W., and A. C. TRAUB. 1968a. "REM Deprivation I: The Effect on Schizophrenic Patients." *Archives of General Psychiatry* 18:287–300.

VOGEL, G. W., and A. C. TRAUB. 1968b. "REM Deprivation II: The Effect on Depressed Patients." *Archives of General Psychiatry* 18:301–311.

VON NEUMANN, J. 1958. *The Computer and the Brain.* New Haven: Yale University Press.

WANG, W., and G. B. RICHERSON, 1999. "Development of Chemosensitivity of Rat Medullary Raphe Neurons." *Neuroscience* 90:1001–1011.

WEHR, T. A. 1992. "A Brain-Warming Function for REM Sleep." *Neuroscience and Biobehavioral Reviews* 16:379–397.

WEIS, J., U. VEBER, J. M. SCHRODER, R. LEMKE, and H. ALTHOFF. 1998. "Phrenic Nerves and Diaphragms in Sudden Infant Death Syndrome." *Forensic Science International* 91:133–146.

WIESEL, T. N. 1982. "Postnatal Development of the Visual Cortex and the Influence of Environment." *Nature* 299:583–591.

WILLIAMS, S. M., E. A. MITCHELL, R. SCRAGG, R. P. K. FORD, I. B. HASSALL, A. W. STEWART, E. M. ALLEN, D. M. O. BECROFT, B. J. TAYLOR, and J. THOMPSON. 1997. "Why is Sudden Infant Death Syndrome More Common at Weekends?" *Archives of Disease in Childhood* 77:415–419.

WILLINGER, M., H. J. HOFFMAN, K. T. WU, J. R. HOU, R. C. KESSLER, S. T. WARD, T. G. KEENS, and M. J. CORWIN. 1998. "Factors Associated with the Transition to Non-Prone Sleep Positions of Infants in the United States—The National Infant Sleep Position Study." *Journal of the American Medical Association* 280:329–335.

WILLINGER, M., L. S. JAMES, and C. CATZ. 1991. "Defining the Sudden Infant Death Syndrome (SIDS): Deliberations of an Expert Panel Convened by the National Institute of Child Health and Human Development." *Pediatric Pathology* 11:677–684.

WILSON, C. A., B. J. TAYLOR, R. M. LIANG, S. M. WILLIAMS, and E. A. MITCHELL. 1994. "Clothing and Bedding and Its Relevance to Sudden Infant Death Syndrome: Further Results from the New Zealand Cot Death Study." *Journal of Paediatrics and Child Health* 30:506–512.

WINSON, J. 1985. *Brain and Psyche: The Biology of the Unconscious.* Garden City, N.Y.: Anchor Press, Doubleday.

WINSON, J. 1990. "The Meaning of Dreams." *Scientific American* 263(5):42–48.

WYATT, R. J., D. H. FRAM, D. J. KUPFER, and F. SYNDER. 1971. "Total Prolonged Drug-Induced REM Sleep Suppression in Anxious-Depressed Patients." *Archives of General Psychiatry* 24:145–155.

YAO, Y., and W. J. FREEMAN. 1990. "Model of Biological Pattern Recognition with Spatially Chaotic Dynamics." *Neural Networks* 3:153–170.

ZIELKE, H. R., R. G. MENY, M. J. O'BRIEN, J. E. SMIALEK, F. KUTLAR, T. H. HUSSMAN, and G. J. DOVER. 1989. "Normal Fetal Hemoglobin Levels in the Sudden Infant Death Syndrome." *New England Journal of Medicine* 321:1359–1364.

Index

acetylcholine, 25, 28, 114
activation-synthesis hypothesis. *See* dreaming
active sleep: in infants, 108, 160. *See also* REM sleep, in infants
adaptation: in neural systems. *See* spurious memory
Alzheimer's disease (AD), 26, 70, 93, 145–146
American Academy of Pediatrics Task Force on Prolonged Infantile Apnea, 163
aminergic-cholinergic (push-pull) cycle: in sleep. *See* dreaming
aminergic neurons/neurotransmitters/system, 26, 53, 125. *See also* noreprinephrine; serotonin
Amit, Daniel, 49, 71, 82
amnesia, 39
amygdala, 19, 21, 26, 40, 43–44, 51, 113–114, 139
animals, 2
  adaptability, 38, 102
  creativity, 102
  dreaming, 110
  experiments. *See experiments*
  memory, 37–38

apnea. *See* SIDS
apparent life-threatening event (ALTE). *See* SIDS
Aristotle, 151
arousal, 22, 25
Aserinsky, Eugene, 108
association: of memories 46, 98. *See also* spurious memory
association cortex, 23, 43–44
associative memory, 46
attention, 22, 45
attractors, 4, 5, 48–55, 76
  as significant cognitive event (decision), 49, 60
  experiments looking for, 50, 54
  for learning, 50
  as memory/recall, 48–49
  pseudoattractors, 60, 67–68
  See also Hopfield model
attractor neural networks (ANN), 49, 55, 62, 66, 71, 135. *See also* Hopfield model
auditory cortex, 23, 43–44
auditory system, 22
Australian Bureau of Statistics, 179
Australian SIDS Foundation. *See* National SIDS Council of Australia

219

autism, 91–92, 95
awake state. *See* wakefulness
awareness. *See* consciousness
axons
   growth cones, 30–31
   thickness and length, 12
   See also neuron, main components; synaptic connections

Babbit, Raymond, 91
basal nucleus (nucleus basalis of Meynert), 26
Bell, Alexander Graham, 128
Berger, Ralph, 131
binding: into complete memory, 44–45, 51
binding problem, 51
Blackmore, Susan, 73–74, 104
blackout catastrophe, 63, 77
Blakemore, Colin, 43
blindness
   dreaming in the blind, 118
   due to cataracts, 33–34
   during critical period, 33
Bohr, Neils, 128
Boulay, George du, 181
brain: insect, 2
brain: human, 1–2, 19
   capabilities and amazing functions, 1–2, 4, 23
   chemical nature of, 3, 17. *See also* chemistry, of brain; neurotransmitters
   computational abilities, 16, 47–48
   cognitive functions, 1, 20, 23
   development of, 5, 8, 29–33. *See also* critical period
   electrical nature of, 3, 17
   hemispheres, 20, 24
   information preprocessing, 13, 44, 64
   long-range excitatory interactions, 45
   network organization, 29, 64
   number of neurons, 2, 11
   number of synaptic connections, 3, 19
   origins of a large human brain, 73
   organs. *See* neuroanatomy
   physical properties, 1, 29–30
   senses, 11
   short-range inhibitory interactions, 45
   storage capacity (memory), 64–65
brain-stem, 19, 21–23, 26–27
   generator of dreams, 7, 22, 113–115
   gliosis. *See* SIDS, pathology
   inhibitor of motor actions during dreaming, 7, 22, 110, 157
   cholinergic injection into, 114, 146
   life-support systems, 24
   rhythms, 21. *See also* rhythms
   See also dreaming; REM sleep
brain *versus* computer, 11, 46–48
   adaptability, 97–98
   computational speed, 47
   creativity, 72, 75–76, 101–102
   memory storage/addressing/recall, 4, 38, 42, 46–47, 75, 134
   noise (in)tolerance, 4, 47
   serial *versus* parallel processing, 47, 134
   storage capacity, 64–65
Broca's area, 23–24

Cajal, Santiago Ramón y, 13, 30
central nervous system (CNS), 11
cerebellum, 12, 19, 21
cerebral cortex (neocortex), 16, 19–24, 26, 42–44, 70, 113
   connections in, 24, 44
   layers, 24
   locations of specific cognitive functions, 23

cerebral cortex (*continued*)
  picture of, *16*
  size and thickness, 20
Changeux, Jean-Pierre, 31–32, 35–36
chaos (chaos theory), 2, 29, 49, 86,
chemistry of the brain, 27, 36
  alterations during dream sleep, 88–89, 117
  alterations with drugs, 27, 52–53, 68, 89, 117
  See also neurotransmitters; drugs; dreaming
chemoreceptors, 158
children
  dreaming, 121, 159
  learning capabilities, 34–35
  neural development, 30, 34–35
  social development, 34–35, 90
chimpanzees, 38, 102
cholinergic neurons/neurotransmitters/system, 26, 53, 114, 123, 125, 145–146
circadian (sleep/wake) rhythm. *See* rhythms
colic, 194, 196
columns: of cerebral cortex, 24, 33, 44
  eye-dominance experiments, 33
  horizontal and vertical line-detection experiments, 34
  in somatosensory cortex, 24
  microcolumns, 44
computers. *See* brain *versus* computer
connectivity: neural. *See* synaptic connections
conscious dreaming. *See* lucid dreaming.
conscious memory. *See* memory, types
consciousness (awareness), 20, 22, 27, 40–41, 45, 53, 115–116
  during dreaming, 53, 115–116, 144.
  See also lucid dreaming

and learning, 40
stream of consciousness, 144
See also self-consciousness
content addressable memory. *See* memory, characteristics; Hopfield model
Cooper, Grahame, 34
coping. *See* memes
corpus callosum, 19, 24–25
corpus striatum, 27
correlation catastrophe, 64. *See also* blackout catastrophe
cortex. *See* cerebral cortex
cot death. *See* SIDS
creativity (creative ideas), 6, 9, 42, 72–74, 86–93, 127, 137
  nature of, 86–88
  copying of, 74, 87. *See also* memes
  in humans, 72–74, 90, 92, 102
  origin (spurious memory), 6, 9, 74–75, 77, 82–85, 90
  versus knowledge, 90–93, 102–103
  See also spurious memory
creativity: generating, 88–90
  after sleep, 127, 137
  by change of environment, 90
  during dreaming, 88, 128–129
  with drugs, 89
crib death. *See* SIDS
Crick, Francis, 8, 69, 85, 88, 105, 122, 127, 133–135, 138, 141, 145, 155
Crick-Mitchison (reverse-learning; unlearning) hypothesis, 8, 105, 121–123, 133–135, 139, 141–144, 155
  absence of learning neurotransmitters, 122–123
  computer simulations of, 122–123, 135, 137–138
  elimination of spurious memories, 8, 105, 122, 135–137

Crick-Mitchison (*continued*)
    equalization of stored memories, 127, 136–137, 147
    excited state (lack of inhibition), 140
    generating spurious memory, 9, 105, 127, 129, 137
    fantasy, reducing, 127
    forgetting, 106, 126–127, 161
    grading memory, 126
    inability to remember dreams, 8, 122
    large amount of REM in infants, 123
    (mild) memory consolidation in, 8, 121–122, 139
    obsession, 122, 127, 139, 142, 148
    partial unlearning, 138
    recurrent dreams, 144
    removal of memories, 69, 125–126, 141–142
    selective synaptic unlearning, 138
    unlearning and learning in alternate cycles, 137
    unsupervised process, 139–140
    weakening of strong memories, 95
critical period, 5, 33–34, 124
    in older children, 34–35
    in infants, 34, 124
    in kittens, 33–34

Dale's law, 17, 25
daydreaming, 88, 115, 128
declarative memory. *See* memory, types
déjà vu, 6, 61, 97. *See also* jamais vu
delay activity experiments, 50, 54
delta (deep) sleep. *See* sleep
Dement, William, 108–110, 153
dendrites
    growth cones, 30, 31
    See also neuron, main components
depression, 26, 146
Descartes, Rene, 20

dolphins, 110, 133, 141
dopamine, 25–28
DNA (dioxynucleic acid), 29–30, 134
dream attractors, 122, 133, 135, 138
dream sleep, 7, 25–26, 108, 134, 155.
    See also REM sleep; dreaming
dreaming, 7, 9, 25–26, 108–110, 113–118, 155
    activation-synthesis hypothesis, 114–115, 133
    brain activity, 118. *See also* lucid dreaming
    aminergic-cholinergic push-pull cycle, 114–115
    blocking of sensory input, 116
    brain-stem-stimulation, 7, 21, 113–115
    consciousness, 53, 115–116
    deactivation of dream action, 7, 115
    dynamics of, 113–118, 133
    different memories than awake state, 123, 139, 149
    eye movements, 117
    function of. *See* REM sleep, functions and theories
    head and body position, 116
    in animals, 110, 120
    in blind, 118
    in infants, 121, 123, 157, 159, 160
    in children, 121
    logic: absence of, 116, 129, 150
    psychoanalysis, 119, 143
    in SIDS, 156, 160, 170. *See also* FMD hypothesis
    See also lucid dreaming; REM sleep; dream content
dream content, 142–144
    amnesia, 122, 142, 144
    association with distress, 118
    association with sex, 119, 131

dream content (*continued*)
  bizarre nature of, 7, 114–115, 143
  concern with more recent memory, 125–126
  concern with personal memory, 7, 119, 143, 157
  delusion, 115
  dream condensation, 143
  emotion, 114
  immobility, 115
  irrelevance of, 7, 119, 143
  old memories, 143
  pain and smell, absence of, 115
  narration, 143–144
  recurring, 117, 144
  red-shaded goggle experiments, 125
  time lapse, 152–153
  See also dreaming; lucid dreaming.
drugs, 27, 146
  effect on brain, 27, 36
  in altered states of mind and creativity, 89
  resulting in different memories, 27, 52–53, 68, 89, 117
dynamical system, 48–49

echidna (Australian anteater), 110
Edison, Thomas, 128
Einstein, Albert, 90, 128
emotion, 21, 27, 40, 51, 53. *See also* amygdala
emotional memory 21. *See* memory, types
epilepsy, 39
episodic memory. *See* memory, types
evolution
  biological, 35, 73, 111
  memetic, 73, 93–94, 102
  neural/synaptic, 5, 32, 35–36

excitatory neurons/neurotransmitters, 16, 26
experiments
  in humans, 39, 111–112
  in neuroscience, 21–22, 33–34, 38, 42–43, 50, 54, 110–112, 120, 157–158, 162,
  in psychology, 38, 41–43, 65, 111, 125
  with animals, 7–8, 21–22, 33–34, 38, 42, 50, 54, 110–112, 114, 120, 157–158, 162
  with computers, 8, 48, 59, 122–123, 135, 137–138
eye-movement desensitization (EMD), 140

fact memory. *See* memory, types
false associations, 93
false memories, 41, 101
fantasy, 26, 127, 139. *See also* spurious memory
fetus, 29–33, 158–159
  heart rate, 188
  neural development, 29–30, 32
  REM sleep, 109, 161
fetal memory dream (FMD) hypothesis, 9–10, 156–157, 161–163, 165, 171–172, 180, 184, 188–189, 194–198
  age at death, exponential decay, 16 161, 168
  age at death, hiatus in first month, 160, 170–171
  association with REM/dream sleep, 157–161, 163, 167, 170
  bed-sharing and room-sharing, 10, 185
  breathing pathways, 157, 161
  breast-feeding, 189, 196

fetal memory dream hypothesis (*cont.*)
  central sleep apnea, 159, 163
  comparison to fetal position, 176–178
  connection with cold weather, 179
  connection with covering, 180–182
  connection to lack of light through eyelids, 180
  connection with warmth, 180, 182
  dream actions, 9, 158, 160–162,
  fetal memory, dreaming of, 9–10, 156–159, 176–177, 179–182, 186, 189
  infant health, 166
  maternal smoking, 184
  pacifier and thumb-sucking, 10, 195
  prevention, sleeping environment, 194
  sleeping position, 10, 175–178, 180
FMD hypothesis: testing the theory further, 191–194
  animal models, 193–194
  fetal environment conditions, 193–198
  infant environment conditions, 194, 195, 198
  monitoring, 191–192
  time of death since start of sleep, 192–193
fixed-point state, 48–49. *See also* attractors
Fontanari, Jose, 98
forgetting: of memory
  biological importance, 69
  by dreaming, 106, 126–127, 141–142
  by decay of neurotransmitter, 127
  in neural network models, 67
  in reality, 68–69
  through memory interference, and spurious memory, 69, 101
  through synaptic growth and pruning, 70
Freud, Sigmund, 8, 118–119, 125, 131, 143–144
frontal cortex, 23
frontal lobe, 23–24

GABA (gamma-aminobutyric acid), 25, 28, 115
generalization. *See* spurious memory; Hopfield model
genes: in neural development, 29–32, 73, 123–124, 159
glutamate (glutamic acid), 25–26, 28
glycine, 27–28
Golgi stain, 13
grandmother cell theory, 42, 84
gray matter: of brain, 11. *See also* neuron
Guilleminault, Christian, 163
Guntheroth, Warren, 162, 198

habituation, 3, 17
Hearne, Keith, 152
Hebb, Donald, 48
Hebbian learning, 50–52, 95, 133
  symmetric form, 56–57
Hebb's hypothesis, 50
heterosynaptic connections, 17
Hillman, Laura, 189
hippocampus, 19, 21, 26, 39–40, 43, 45, 51, 113, 120–121, 138–139, 146
  lamellae, 70
  storage capacity, 70
  transfer of information to neocortex, 39, 70, 120–121, 126, 138
histamine, 27–28
H. M. (famous patient), 21, 39, 70
Hobson, J. Allan, 22, 113–114, 116, 125, 133, 155, 171

Hopfield, John, 48, 55–56, 71
Hopfield model, 5, 55–66, 78, 101
  ±1 neural activity models, 56, 76, 78
  0/1 neural activity models, 65–66, 76, 84
  basin of attraction, 58, 61, 68, 84, 87, 137
  coding level, 66, 84
  connectivity, 56, 59, 66–68, 85
  content-addressable memory, 55, 58, 60
  convergence to attractor, definite, 60
  convergence rate, 6, 59, 61. *See also* déjà vu
  distributed storage of memory, 61–62
  double (synaptic and neural) dynamics, 52
  dynamics (updating procedures), 59–60
  earlier models, 60
  energy states, minimum, 57, 60–62, 76
  energy minimization process, 55, 60–61. *See also* memory landscape
  error/noise tolerance, 58–59, 61–62
  generalization and categorization, 98–100
  learning rule, symmetric Hebbian, 56–57
  learning rules, other, 65
  loading level, 82
  models with more realistic features, 60, 65, 83–85. *See also* ANN
  network update cycle, 59, 61
  overlapping storage of memory, 61–62
  recall of memory, from incomplete input, 58–59, 60–62
  recognition, 49, 97
  recurrent (totally connected), 59–60
  simulations, 59, 135, 137. *See also* experiments
  spin-glass states, 62, 82, 95, 97. *See also* spurious memory
  spurious memories, 61–64. *See also* spurious memory
  spurious memory, example of, 62–63, 78, 80–82
  static memories, 54. *See also* attractors
  stored memories, as patterns, 56–57, 61–62
  stored memory, example of, 57, 78–79
  storage capacity, 57, 64, 66
  synaptic weights, 57. *See also* synaptic efficacies
  temporal memories, 53–55, 99–100
  threshold, 59, 66, 84
  unsupervised storage (autonomous function), 51–52, 57. *See also* spurious memories, in new learning
Howe, Elias, 128
Hubel, David, 33
Huntington's Chorea, 25
hypothalamus, 20–22

inhibitory neurons/neurotransmitters, 16, 25–27
infant, 8, 29
  different types of REM sleep, 121, 160, 170
  development of dreaming apparatus, 171
  dreaming REM sleep, 8, 121, 157, 159
  heart rate, 188
  reduced inhibition of dream actions, 160

infant (*continued*)
  REM sleep, 106, 108–109, 121
  neural development, 8, 29–33
intelligence, 92
Internet, 37, 73

jamais vu, 6, 97. *See also* déjà vu
jet lag, 23
Jouvet, Michael, 111, 131
Jung, Karl, 118

Kekulé, Friedrich, 88, 128
Kleitman, Nathaniel, 108
knowledge, 72–73, 90–93, 102–103
  limit on human understanding, 103–104

LaBerge, Steven, 151–153, 155
Landau, Lev, 91
Landauer, Thomas, 65
language, 20, 25, 34, 37, 73
Lashley, Karl, 42–42
learning, 3, 25–27, 50–53
  associated neurochemistry, 26–27, 51, 114
  attractors and reverberations, 51
  consciousness, connection with, 40
  emotions, connection with, 40, 51
  flexibility in, 35–36
  neurobiological mechanics, 51–52
  rehearsal and repetition, 40, 52
  relearning, 96
  spurious attractors, 6, 9, 52, 68, 95–96
  unsupervised dynamic learning, 51–52, 57
  See also norepinephrine; locus coeruleus
limbic system, 26, 113, 139
limit cycle, 48–49

Little, William, 60
lobotomy: pre-frontal, 24
locus coeruleus, 22, 25, 51, 114
  vast connectivity to neorcortex, 22, 114
  See also norepinephrine
Loewi, Otto, 26, 88, 128
long-term memory. *See* memory, types
LSD (lysergic acid diethylamide), 26, 89, 143. *See also* serotonin
lucid dreaming, 9, 89, 118, 144, 149–155, 157, 176
  actions during, 9, 157
  brain activity, 118, 153, 154
  communication with laboratory, and dream reports, 9, 152
  consciousness, 149–151
  control over dream content, 9, 149
  dreamlight machine, 151
  frequency, 150–151
  lapsed time, 152–153
  mnemonic induction of lucid dreams (MILD), 151
  specific dreams, examples of, 9, 150, 153–154, 157

mammals, 20, 28, 110, 133, 141
Mandelbrot, Benoit, 29
marijuana, 89
mathematical models, 5, 41, 72, 74, 134. *See also* Hopfield model; ANN
Maurice, David, 131
Maury, Alfred, 152–153
McCarley, Robert, 22, 113–114, 133
medium-term memory. *See* memory, types
medulla, 19, 110
memes, 72–74, 88, 93–94, 104
  and creativity, 74, 93–94
  copying, 72–74, 87

memes (*continued*)
  imperfect transmission of, 74, 87–88, 94
  evolution of (memetic evolution), 73, 93–94, 102
  origin of language, 73
  origin of big human brain, 73
memory: neural, 3–5, 11, 18–21, 23, 26–27, 37–47, 71, 75,
  definition of, 3, 37
  biological importance/usefulness, 3, 37
  consciousness, with, 40. *See also* memory, types; learning
  consolidation, 40, 112, 120–122
  drugs, and, 27, 52–53, 68, 89,
  grandmother cell hypothesis, 42, 84
  localized storage in modules, 44, 84,
  obsession, 95
  recall/retrieval of, 4–5, 44–45, 48–48. *See also* complete memory
  recalling from cues, 4–5, 43, 45–46, 69, 75
  recognition, 4, 46, 48–49, 97
  search for storage location (engram), 42–43
memory: characteristics
  associative, 46, 98
  content addressable, 5, 46, 75, 87,
  distributed storage of, 4–5, 41–43, 75, 134
  decay of, 47. *See also* forgetting
  duration of, 41, 69. *See also* memory, types
  improving memory, by making associations, 46
  limitation to 7±2 items in short term memory, 41
  overlapping storage of, 4–5, 41, 72, 75
  robustness of, 45, 47, 72, 75

memory: in
  immune system, 37
  nervous system, 37–38. *See also* memory
  plants, 37–38
  neural network models. *See* Hopfield model; ANN
memory: types, 38
  declarative/explicit/conscious, 40, 101, 126
  emotional memory, 21, 40, 44
  episodic/personal, 40, 43
  long-term, 21, 26, 39–41
  medium-term, 21, 26, 39, 41
  motor/skill, 21, 39–40
  permanent/hardwired, 41
  semantic/fact, 40
  short-term, 21, 26, 41, 146
  temporal/dynamic, 53–55, 65, 99–100
  non-declarative/implicit/unconscious, 40, 101, 126
  unconscious/subliminal form of short-term, 41
  working, 41
  weak/irrelevant/trivial memory, 125–126
memory landscape (memoryscape), 61–62, 75–77, 95–96, 100, 135–136, 147–148
Mendeleyev, Dmitry, 88, 128
Milner, Brenda, 39
mind: human, 1, 20, 25, 43, 52–53, 73, 89–90, 165. *See also* self
Mitchison, Graeme, 8, 69, 85, 88, 105, 122, 127, 133–135, 141, 145, 155
Miyashita, Yasushi, 50, 54
modules: neural. *See* neural networks, biological
mood, 27, 53

motor cortex, 23
motor neurons. *See* neuron, types
motor output, 21
motor/skill memory. *See* memory, types
Mountcastle, Vernon, 24
multiple personality disorder, 53

National SIDS Council of Australia, 189, 194
natural selection, 35. *See also* evolution, biological
navigation memory, 43
neocortex. *See* cerebral cortex
nervous system, 11. *See also* CNS; brain; neural networks, biological
Neumann, John von, 64–65
neural networks: biological, 13–15
   collective behavior, in memory recall, 44–45, 47
   complex connectivity, 12, 13
   real photographs of, 15–16
   modules, 5, 24, 44–45, 49, 84
   noise tolerance, 47, 50
neural networks: models, 5–6. *See also* Hopfield model; ANN
neuroanatomy, 19. *See also under* amygdala; brain-stem; cerebellum; cerebral cortex; corpus callosum; hippocampus; hypothalamus; locus coeruleus; medulla; neocortex; pons; raphe nuclei, REM-on cells; spinal cord; thalamus
neurochemistry of brain. *See* chemistry of brain; neurotransmitters
neuron, 2, 12–14, 16, 25, 29
neuron: main components
   axon, 12–14, 16–17
   dendrite, 12–14, 16–18
   soma, 12–14, 16–18

   See also synapse
neuron: properties
   chemical nature of, 17–18. *See also* neurotransmitters
   collective behavior, 43, 45, 47, 64. *See also* neural networks, biological
   communication means/channels, 2, 11, 14–18. *See also* synapse; neural networks, biological
   electrical nature of, 3, 13, 17–18
   excitation of, 3, 12, 14–15
   input, 2
   output, 3
   refractory period, 16
   threshold (potential), 12, 15
neuron: types
   excitatory neurons, 12, 16
   inhibitory neurons, 16
   interneurons, 12–13
   motor neurons, 12–13
   Purkinje cells, 12
   pyramidal cells, 12, 16, 45
   sensory neurons, 11–12
   stellate cells, 12
neuron doctrine, 13
neuronal death, 29–32
neuroreceptors, 17
neurotransmitters, 3, 4, 17–18, 21, 25–27, 50–52, 67
   See also under acetylcholine; dopamine; GABA; glycine; glutamate, histamine; norepinephrine; serotonin
   See also synapse, vesicles
new learning. *See* spurious memory, in new learning
newborns, 29, 119
   dreaming, 160, 171
   gasping reflex mechanism, 170

newborns (*continued*)
    neural development, 170–171
    REM sleep, 119, 157
nitric oxide (NO), 18, 51
non-declarative memory. *See* memory, types
norepinephrine, 25–26, 28, 51, 53, 114–115
    activity during wakefulness, 25, 114
    (in)activity during dream sleep, 25, 114–115
    association with attention and learning, 25–26, 114, 133
    association with consciousness, 53, 116
    See also locus coeruleus
NREM sleep. *See* sleep
NREM/REM cycle. *See* rhythms
nuclei, 22–23, 44

obsession, 95, 121–122, 139, 142, 148. *See also* CM hypothesis
occipital lobe, 23
oculomotor neurons, 117, 140
optical chiasm, 23
overloading catastrophe, 64–65, 67, 77, 83. *See also* blackout catastrophe

palimpsest models, 67
    storage capacity, 67
    with bounds, 67, 83
    with synaptic decay, 67
    See also Hopfield model
paradoxical sleep, 108. *See also* REM sleep
parasitic modes, 122, 135, 139. *See also* spurious memory
parietal lobe, 23
Parisi, Giorgio, 85

Parkinson's disease, 26
Pavlov, Ivan, 46
perception, 45, 52–53
personal memory. *See* memory, types
personality, 23, 25, 53. *See also* mind
phonetics: during critical period, 34
pineal gland, 20
Poincaré, Henri, 86, 89–90
pons, 19, 22, 110
postsynaptic cell, 17–18
posttraumatic stress disorder, 140
presynaptic cell, 17–18
prozac, 26
psychoanalysis. *See* dreaming
psychological theories, of dreaming. *See* REM sleep
psychosis, 144–145. *See also* schizophrenia
Purkinje neurons. *See* neuron, types
pyramidal neurons. *See* neuron, types

random (noisy) signals, from brainstem, 7–8, 113–115, 133. *See also* REM-on cells; dreaming
raphe nuclei, 26. *See also* serotonin
REM/NREM cycle. *See* rhythms
REM (rapid eye movement) sleep, 7, 22, 26, 105–112, 133, 138, 144–146, 155, 157–158
    activity of brain, 105, 108, 133, 158
    alertness, 106–107
    amount of, 8–9, 109, 124, 157, 159
    association with dreaming, 7, 22, 108–109, 121, 157
    biological importance, 7, 110–112, 194, 197
    brain-stem stimulation, 7, 113–114, 157–158
    cerebral blood flow, 7, 108

REM sleep (*continued*)
  consolidation of memory, 112, 120–121
  deprivation experiments in animals, 111–112
  deprivation in humans, 111–112, 145
  desynchronized electrical activity of brain, 7. *See also* EEG
  disconnection from body and environment, 7, 116, 152, 154, 157
  dual purpose in infants, 121
  EEG, 107–108
  enhancement by microinjection, 114, 146
  enhancement with drugs, 146
  experiments with cats, 7–8, 22, 110, 157–158
  immobility, 22, 108
  mobility in infants, 108, 131. *See also* active sleep
  muscle-tone relaxation, 108
  pattern of during sleep, 106–107
  problems with REM loss or deficit, 144–146
  psychotic behavior and deprivation, 111, 145
  REM-rebound, 7, 110
  REM latency, 111
  similarity to awake state, 108, 133
  suppression with drugs, 145
  See also dreaming
REM sleep (and dreaming): functions and theories, 7, 118–142
  acting out future scenarios, 130
  converting declarative to non-declarative memories, 126
  creative solutions, 128–129
  entertainment, 130
  erection (male), 131
  eyes wet, 131
  forgetting, 106, 126–127, 141–142
  generating more spurious memory (roughening memoryscape), 9, 105, 127, 129, 146–149
  grading of memory, 126
  head size, 110, 141
  hippocampus teaching neocortex, 120–121, 126, 138
  memory consolidation, 8, 119–121
  motor skills in infants, 131–132
  multifunctional, 132
  no function whatsoever, 132
  on alert, 129
  ontogenetic, 8, 123–124, 159
  pruning (and developmental), 124
  psychological theories, 8, 118–119, 142, 144
  rehearsal/relearning, 8, 105, 120, 139
  removal of trivial memories, 125–126, 141–142
  replenishment of aminergic neurotransmitter, 125
  reprocessing daily information, 125–126
  reprocessing general memory, 108
  warm brain, 130
  uncommon behavior, 131
  visual perception, 131
REM sleep: in
  adults, 8–9, 109, 124, 157, 159
  children, 119, 124
  dolphins, 110, 133, 141
  infants, 8–9, 106, 108–109, 123–124, 131–132, 157, 159, 160
  fetus, 109, 119
  mammals, 110
  newborns, 119
REM-off cells, 114

REM-on cells, 22, 26, 113, 116, 140
reverberations, 48, 50. *See also* attractors.
reverse-learning, 8, 121–122, 133–142, 161. *See also* CM hypothesis
rhythms: of the brain, 21
    circadian (sleep/wake), 21–23. *See also* jet lag
    rest-activity, 115
    sleep/dream (NREM/REM, SW/REM) cycle, 21, 106–107, 111, 114–115, 129–130. *See also* dreaming, aminergic-cholinergic cycle
Roffwarg, Howard, 109, 124

schizophrenia, 27, 93
Schwartz, Peter, 188
self, 73–74
self-consciousness (self-awareness), 53
    in autistic savants, 91
    in chimpanzees, 38
self-organization: in neural systems, 30–32
semantic memory. *See* memory, types
sensory input 20–23. *See also* thalamus
sensory neurons. *See* neurons, types
serotonin, 25, 26, 28, 53, 114–115
    association with attention and learning, 26, 133
    during dreaming, 114–115, 143
Shaw, Gordon, 60
short-term memory. *See* memory, types.
sleep, 6, 105–106, 155
    alertness during, 106–107
    alternate SW/REM (NREM/REM) cycles. *See* rhythms
    deep (delta) sleep, 106
    electroencephalograms, 106–107

    function of sleep, 6, 105
    hypnograms, 107
    in infants, 106, 109
    in children, 106
    in adults, 106
    in newborns, 106
    NREM sleep, 106, 163
    stage 4 rebound, 111
    stages of sleep, 106–107
    SW sleep, 106, 111–112, 138
    See also REM sleep
sleep/dream cycle. *See* rhythms
sleep/wake cycle. *See* rhythms
social behaviour, 38, 72–74, 90
society: human, 1, 37, 72–73, 94–95, 102
soma. *See* neuron, main components
somatosensory cortex, 23, 44
somatosensory system, 22
speech. *See* Broca's area
Sperry, Roger, 31
spinal cord, 11–12, 19, 21–22, 27
spin-glass states. *See* spurious memory
spinocerebellar degeneration, 132
spurious memory, 5, 8, 41, 62–64, 75–87, 94–103, 121–122, 127, 134–137, 143
    elimination of during REM sleep, 8, 105, 122, 135–136
    examples of, 62–63, 78, 80–82
    exponential growth in Hopfield model, 63–64, 77
    in adaptation, 6, 9, 36, 97–98, 127
    in association, 6, 46, 98, 147
    in creativity, 6, 9, 68, 74–75, 77, 82–85, 87, 127, 129, 134, 137. *See also* creativity
    in forgetting, 101. *See also* forgetting

spurious memory (*continued*)
  in generalization and categorization, 6, 68, 98–100, 134
  in new learning, 6, 9, 52, 68, 95–96, 134, 147–148
  in process of discovery, 103
  in temporal memory, 55, 99–100
  in thinking and planning, 9, 101
  interference with recall of stored memory, 5, 62, 65, 121–122, 134
  making mistakes, 6, 74, 87–88, 98
  more realistic models in, 68, 77, 83–85
  need to control, 62–65, 76–77, 81
  non-symmetric states, 82
  number in Hopfield model, 80–81
  origin of, 5, 62–63, 74–77
  psychological problems, 93
  spin-glass states, 62, 82, 95, 97
  symmetric states, 78, 80–81
  symmetry amongst spurious memories, 77, 82,
  3-mixture and 5-mixture states, 80–81
  See also blackout catastrophe; creativity; memory landscape
spurious memories:
    control/reduction/destabilization of (in Hopfield model), 66–67, 83–85
    by unlearning, 66, 85, 122, 134
    by less overlap of memory, 66, 84–85
    by reduced coding level, 84
    by increasing threshold, 84–85
    by introducing noise, 84
    by bounded synaptic efficacies, 67, 83
    by asymmetric connectivity/action, 85
    by storing memories with unequal intensity, 85
Snyder, Frederick, 129

stages of sleep. *See* sleep
static memory. *See* memory, types
Steinschneider, Alfred, 67
stellate neurons. *See* neuron, types
strange attractors, 48–49, 85
  used as cognitive significance, 49
  See also chaos
Stevenson, Robert Louis, 129
subconscious (mind), 90, 143
sudden infant death syndrome (SIDS), 9, 10, 141, 156, 168
  in bible, 167
  climatic link, 179–182
  cultural link, 167–168, 174, 179,
  deaths on monitors, 188
  definition of, 168
  diagnosis, 10, 156, 164–165
  during REM/dream sleep. *See* FMD hypothesis
  during sleep, 166–167, 192
  epidemiology, 156, 164, 171
  gasping effectiveness, 174
  good health, in, 165
  in dolphins, 141
  in 'awake' babies, 166–167
  long QT interval, 188,
  medical mystery of, 9, 156, 164–165
  number of deaths/mortality rates, 9, 156, 165
  other causes of infant death, compared with, 165, 169
  sleep organization, heart-rate and breathing pattern activity, 164, 187–188
  suffocation murders/ infanticides, 164–165, 187, 190
  time of death during sleep (early morning), 167, 186
  trigger mechanism, 157, 159, 191

SIDS: age at death distribution, 121, 168–169, 181
  exponential decay, 168–169, 171
  hiatus (low incidence) in first month, 160–161, 169–171
  premature infants, for, 171
  See also FMD hypothesis
SIDS: apnea, 163, 175
  central, 159, 162–163
  obstructive, 162–163
  sleep, 162
SIDS: demographics
  in Australia, 174, 179
  in Canada, 179
  in Denmark, 173, 177
  in Finland, 168, 174, 179
  in Hong Kong, 168, 174
  in Japan, 168, 174, 179
  in New Zealand, 168, 174, 179
  in Norway, 173, 177
  in Sweden, 168, 173–174, 177, 179
  in the Netherlands, 172
  in United Kingdom, 174, 181
  in United States, 156, 168, 174, 179, 182
  in the world, 166
SIDS: matters not likely related to SIDS
  alcohol, 183
  ALTE, 163, 190. *See also* SIDS, and apnea
  choking and vomiting, 189, 194
  cots, 167
  genetics link, 190
  immunization, 189
  infection and colds, 189
  near-miss SIDS, 163, 190
  pathogen, 166
  poison, 166

  siblings, 190
  suffocation, 165, 173–174
SIDS: pathology, 163–164, 186–187
  brain-stem gliosis, 183, 186–187
  fetal hemoglobin, 187
  petechiae and brown fat, 186
  neurotransmitter/neuroreceptor alterations, 186–187
  immaturity of brain-stem, 186
SIDS: possible main causes of death,
  respiratory cessation 159, 163, 186–187, 191. *See also* FMD hypothesis
  heart failure, 163
  suffocation, 173, 175
SIDS, prevention
  back-to-sleep campaign, 173, 195. *See also* SIDS, sleeping position
  breast feeding, 167, 186, 189
  maternal smoking, reducing, 195
  monitoring, 188, 191–192
  noises, 195
  pacifier sucking, 10, 195
  recommendations by SIDS foundations 194–195
  reduced covering (especially of head), 180, 195
  reminder of being born, 185, 194–195
  room-sharing, 10, 185, 195
  sleeping environment, 194–196
  temperature comfort, 195
  tucking baby in tight, 180, 194
  See also FMD hypothesis
SIDS: reflexes
  chemoreflexes, 158–159, 191
  arousal from sleep, 158–159, 178
  simultaneous failure of all reflexes, 158–159, 163, 191
  development of, 160–161
SIDS: risk factors, 10, 156, 171–172

SIDS: risk factors (*continued*)
  apathy and education factor, 183–184, 186
  bed-sharing, 10, 178, 185, 196
  cold climatic regions, 179
  cold seasons, 179, 181
  combined risk factors leading to enhanced risk, 177, 180
  covering, excessive bed, 179–180
  covering baby's head, 180
  drugs, 183–184
  fetal environmental conditions, 194, 196–197
  home heating, poor, 179, 182, 184
  interrelation of risk factors, 171, 177, 180, 182–184
  low birth weight, 171–172, 184
  male *versus* female, 185
  maternal smoking during pregnancy, 171–172, 182–184, 195
  other (less important) risk factors, 184–185
  overheating of infant, 175, 180–181
  premature infants, 177, 180, 184
  sleeping environment, 179
  socioeconomic status, 171–172, 182–184, 186
  soft bedding, 186
  thumb sucking, 10, 195
  sweating, 181
  unimportant (no direct impact), 172
  weekends *versus* weekdays, 185
  See also FMD hypothesis
SIDS: sleeping position, 172–178
  history, 172–174
  prone (face-down)sleeping position, 10, 161–162, 165–166, 172–177, 180, 182, 186, 194–195
  relative risk, 173–174, 194
  reduction of incidence through sleeping position, 10, 161, 165, 172–173, 182, 195
  side sleeping position, 173–174, 177, 194
  supine (face-up) sleeping position, 10, 161, 172, 173–178, 194
  higher risk of prone position for younger infants, 177, 180
  higher risk of prone position for premature and low birth weight, 177
  See also FMD hypothesis
SIDS: theories
  accidental suffocation, 167, 173
  CO_2 rebreathing/poisoning, 175
  dive reflex, 162
  dreaming. *See* FMD hypothesis
  facial overheating, 181
  infection/pathogenic/poisoning, 166–167, 189
  reflex (brain-stem) immaturity, 161, 168, 170, 183
  restriction of blood flow in neck, 174–175
  thermal stress, 175, 181, 184
suprachiasmatic nucleus, 22–23
survival, 3, 11, 30, 74, 110, 120, 141
SW sleep. *See* sleep
SW/REM cycle. *See* rhythms
synapse
  bouton, 16–18
  cleft/gap, 3, 14, 17–19, 27
  size of synaptic gap, 17
  vesicles, 17–18, 26, 50–51, 67. *See also* neurotransmitters
synaptic connections, in human brain
  and brain weight, 29–30
  rate of growth, 29–30, 70
  random nature of, 29–30, 97

synaptic connections (*continued*)
  in womb, 30
  growth cones, 30–31
  growth, transient redundancy, 29–32, 35–36, 97, 124
  plasticity, 35
  pruning, 29–32, 35–36, 70, 97, 124, 159
  specific connections and genes, 31, 123–124
  number in adult brain 3, 19
  number in infant brain, 29–30
synaptic, efficacies/weights, 50
  bounds, 67, 69, 83
  decay of, 67
  quantization of, 52, 67, 83
  See also palimpsest models; learning
synchronization in neural systems, 45, 55. *See also* binding

temporal cortex, 23, 39, 50
temporal lobe, 23
temporal memory. *See* memory, types; spurious memory; Hopfield model
thalamus, 19–20, 22, 45, 51, 113–114
  nuclei, 22
  intricate connections to neocortex, 20, 22
  reciprocal connections from neocortex, 22

theta rhythm, 120
thinking process, 86. *See also* spurious memory
Tholey, Paul, 151
threshold, neural. *See* neuron, properties

unconscious memory. *See* memory, types
unconscious (mind), 41, 90, 144
unlearning, 8, 26, 121–122, 133–134, 141. *See also* reverse-learning; CM hypothesis
U.S. National Institute of Child Health and Human Development (NICHD), 168

vertebrates, 20, 31
vesicles, synaptic. *See* synapse
visual cortex, 23, 43–44
visual system, 22

wakefulness (waking consciousness), 22, 106–107, 109, 114–115
Wallace, Alfred, 35
white matter: of brain, 11, 24. *See also* axons; dendrites; corpus callosum
Wiesel, Torsten, 33
Winson, Jonathan, 105, 120, 138
working memory. *See* memory, types
Worsley, Alan, 152

About the Author

GEORGE CHRISTOS was born in Perth, Western Australia. He received his D.Phil. from Oxford University in elementary particle physics in 1981. From 1985 to 1987, he was a QE II Fellow. Since 1991, Christos has been interested in understanding how the brain works, and how memory is processed. He teaches in the Department of Mathematics and Statistics at Curtin University of Technology in Perth.